SCIENCE *of*

UNDERSTAND THE ANATOMY AND PHYSIOLOGY TO TRANSFORM YOUR BODY

HIIT

HIGH INTENSITY INTERVAL TRAINING

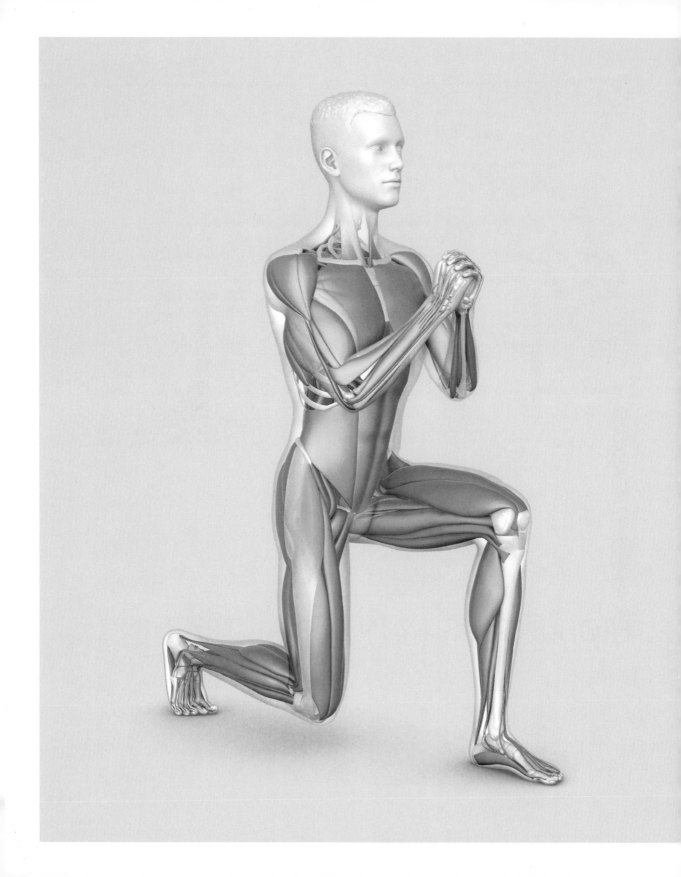

SCIENCE *of*

UNDERSTAND THE ANATOMY AND PHYSIOLOGY TO TRANSFORM YOUR BODY

HIIT

HIGH INTENSITY INTERVAL TRAINING

Ingrid S Clay

DK | Penguin Random House

Project Art Editor Amy Child
Project Editor Susan McKeever
Designer Alison Gardner
Senior Editor Alastair Laing
Senior Designer Barbara Zuniga
US Editor Kayla Dugger
Jacket Designer Amy Cox
Jacket Coordinator Jasmin Lennie
Production Editor David Almond
Senior Production Controller
Luca Bazzoli
Managing Editor Dawn Henderson
Design Manager Marianne Markham
Art Director Maxine Pedliham
Publishing Director Katie Cowan

Illustrations Arran Lewis

First American edition, 2022
Published in the United States by
DK Publishing, 1450 Broadway,
Suite 801, New York, NY 10018

Text copyright © Ingrid S Clay 2022
Copyright © 2022
Dorling Kindersley Limited
DK, a Division of Penguin
Random House LLC
21 22 23 10 9 8 7 6 5 4 3 2 1
001–326788–Jan/2022

Published in Great Britain by Dorling
Kindersley Limited.

A catalog record for this book
is available from the Library of Congress.
ISBN 978-0-7440-5128-5

Printed and bound in China

For the curious
www.dk.com

MIX
Paper from
responsible sources
FSC™ C018179

This book was made with Forest
Stewardship Council ™ certified
paper – one small step in DK's
commitment to a sustainable
future. For more information go to
www.dk.com/our-green-pledge.

CONTENTS

Introduction   06

PHYSIOLOGY OF **HIIT**

The benefits of HIIT   10

Powering your HIIT workout   12

Improving your cardio fitness   14

The HIIT "afterburn"   16

How muscles work   18

How HIIT promotes muscle growth   20

Muscular anatomy   22

HIIT training and the brain   24

Feeding your HIIT   26

HIIT EXERCISES

Introduction to the exercises   30

Terminology guide   32

CORE EXERCISES

High plank to low plank   36
   Variations   38

Swim plank   40

Mountain climber   42
   Variations   44

Bear plank   46
   Variations   48

Sit-up   50

Crunch   52
   Variations   54

Transverse abdominal ball crunch   56

V-up   58
   Variations   60

UPPER BODY EXERCISES

Push-up   64
   Variations   66

Overhead triceps extension   68
   Variations   70

Dumbbell biceps curl   72
   Variations   74

Dumbbell front raise   76
   Variations   78

Dumbbell lateral raise   80

Military shoulder press   82
   Variations   84

Dumbbell rear deltoid fly   86
   Variations   88

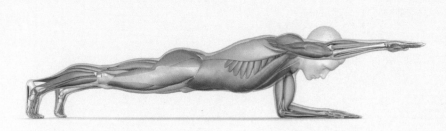

Dumbbell bench press **90**

Dumbbell chest fly **92**

LOWER BODY EXERCISES

Squat **96**
    Variations **98**

Right and left split squat **100**
    Variations **102**

Crab walk **104**

Alternating snatch **106**

Alternating lateral lunge **108**
    Variations **110**

Calf raise **112**

Step-up with dumbbells **114**

Alternating toe tap **116**

Single-leg deadlift **118**

Glute bridge **120**
    Variations **122**

PLYOMETRIC EXERCISES

Skater **126**

High knee **128**
    Variations **130**

Squat jump **132**
    Variations **134**

Tuck jump **136**

Box jump **138**

Single-leg forward jump **140**

Football up and down **142**

Burpee **146**

Bear crawl **150**

TOTAL BODY EXERCISES

Jack press **154**

Push-up and squat **156**

Push-up and tuck jump **160**

Bear plank and push-up **164**

High plank, ankle tap,
    and push-up **168**

B boy power kicks **172**

Military press and overhead
    triceps extension **174**

Bent-over row and
    hammer curl **178**

Rear deltoid fly and
    triceps kickback **180**

Sumo squat and hammer
    concentration curl **184**

HIIT TRAINING

Getting started **188**

Planning your training **190**

Following and creating
    your own routine **192**

Weekly training planner **194**

Warming up
    and cooling down **196**

Workout routines **198**
    Beginner **199**
    Intermediate **202**
    Advanced **206**

Glossary **210**

Index **214**

Bibliography **222**

Acknowledgments **223**

About the author **224**

INTRODUCTION

When it comes to burning fat and toning the body, high-intensity interval training, more commonly known as HIIT, has always been at the forefront of training regimes. It's easy to see why—HIIT-based exercises combine dynamic bursts of cardiovascular activity with resistance-based strength-building movements, and you can be completely done with your routine in as little as 20 minutes. HIIT exercises alternate between short periods of intense anaerobic exercises and a less intense recovery period. This book aims to explain why HIIT-based training is effective by looking at the science behind the exercises. It will also teach you, through detailed anatomical illustrations and instructions, how to perform these exercises correctly—whether you are an absolute beginner or a fitness fanatic. The beauty of HIIT is that you can add it to your current training program, you can do it at home or in the gym, and it doesn't require a huge time commitment. You'll find that the information in this book will help you put together an effective HIIT workout that suits you and will build your knowledge and confidence in order to execute it correctly.

Why HIIT?

The exercises in this book focus primarily on building cardiovascular strength and endurance, but there are many more benefits to HIIT-based workouts. The nature of a HIIT workout—intense bursts of exercise in a short time—raises your metabolic rate for up to 24 hours after your training session is finished, essentially turning you into a fat-burning machine! We will go into the many benefits of HIIT on pp.10–11, but here are just a few to whet your appetite:

- **Burns calories more quickly** with the short, sharp approach, compared to other types of training.
- **Boosts the body's cardiovascular health** and reduces blood pressure.
- **Aids in reducing anxiety and depression**.
- **Improves athletic performance** by raising your anaerobic efficiency, increasing your VO2 max score, and building and maintaining muscle (see pp.10–21).

How This Book Works

The first section of the book discusses human physiology, explaining the science behind how HIIT improves your cardiovascular fitness, raises your metabolism and fat-burning rate, and builds and tones your muscles. I will also walk you through how best to fuel your workouts with the correct macronutrients—protein, fats, and carbohydrates—needed not only to effectively perform your workouts, but also to achieve your desired goals.

The main section of this book is devoted to a comprehensive collection of HIIT exercises that target

> # *HIIT workouts are short, you can do them **anywhere**, and you'll **burn more** fat in less time than with any other workout.*

different parts of the body, along with modifications and variations for various fitness levels. These exercises come with detailed notes on how to execute them with the proper form, as well as identifying common mistakes made and cautionary advice on how to avoid injuries. Each illustration of the various steps of an exercise is clearly annotated with precise form and posture directions. The book finishes with a selection of easy-to-follow routines geared toward beginner, intermediate, or advanced level.

Science of HIIT is a fantastic starting point for anyone just getting started in fitness, new to HIIT, or for those seasoned athletes looking to give their current training regime a boost. Whether your goal is to create a comprehensive and personalized training program, gain a better understanding of the mechanics behind HIIT exercises, or simply lose weight and tone the body, this book covers it all—it will be your new best friend! As you progress through the exercises and become accustomed to them, you can increase their intensity and duration, making this book a relevant "bible" for years to come.

Ingrid S. Clay
www.ingridsclay.com

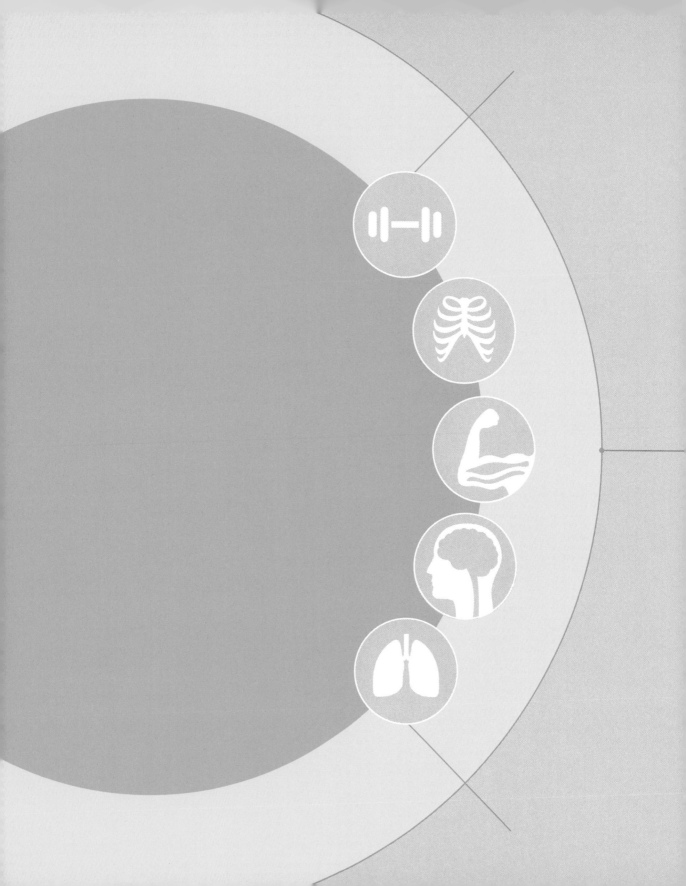

PHYSIOLOGY OF **HIIT**

HIIT training requires you to push yourself as hard as you can during the "working" intervals of your routine, while the rest periods (the intervals) allow you to briefly recharge. Find out more about how this impacts your physiology—your muscles, your bodily systems, and how you process nutrients. Learn also about how HIIT can benefit brain function.

THE BENEFITS OF HIIT

During a HIIT workout, your body shifts between bursts of intense physical activity—comprising a mix of strength exercises and cardio—and short periods of recovery. What benefits does this alternating mode of training bring to health and fitness?

THE DIFFERENCE IN HIIT

Compared to a jog around the park, a HIIT workout is fast and furious; if undertaken with maximum effort, as little as 10 minutes is all that's required. Yet studies suggest that those minutes of HIIT training are far more effective at burning calories and fat than hours of moderate- to low-intensity steady cardio. The reason for this is that a HIIT workout keeps the body guessing. During extended periods of cardio performed at the same pace with the same effort, the body is able to adapt and reach a metabolic state designed to conserve energy. By contrast, because HIIT causes your heart rate and energy output to fluctuate throughout an exercise session, the body fails to find this steady state and its calorie-burning energy needs remain at a higher rate. What's more, this raised metabolic state continues for much longer during postexercise recovery (see pp.16–17).

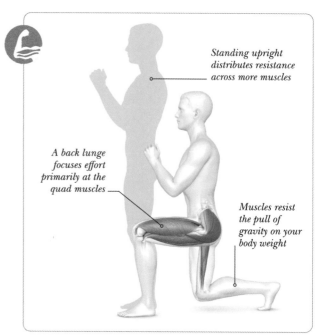

Standing upright distributes resistance across more muscles

A back lunge focuses effort primarily at the quad muscles

Muscles resist the pull of gravity on your body weight

Muscle fibers include filaments that contract to produce movement

Energy release is linked to blood flow and mitochondria (see p.15)

Body weight resistance

The primary focus of physical effort in almost all HIIT exercises lies in body weight resistance—that is, engaging your muscles to resist the force of gravity acting on the weight of your body. The upshot of this is that HIIT requires very little equipment and can be performed at any time and almost anywhere—except maybe on the moon!

Raised metabolism

The body's metabolic system provides the energy needed to power muscle contractions during HIIT. The all-out, stop-start nature of HIIT helps keep your metabolism at a raised state for longer than continuous low-intensity exercise (see pp.12–17).

Immune support

Exercise contributes to overall health, which ultimately benefits the immune system. Scientists have various theories making a more direct link, but research is so far inconclusive. One theory is that increased flow of blood and lymph fluid from exercise improves the circulation of immune cells around the body. Another is that by reducing inflammation, exercise supports immune function that is compromised by states of chronic inflammation. Exercise has also been proven to reduce mental stress, which has an adverse effect on the immune system.

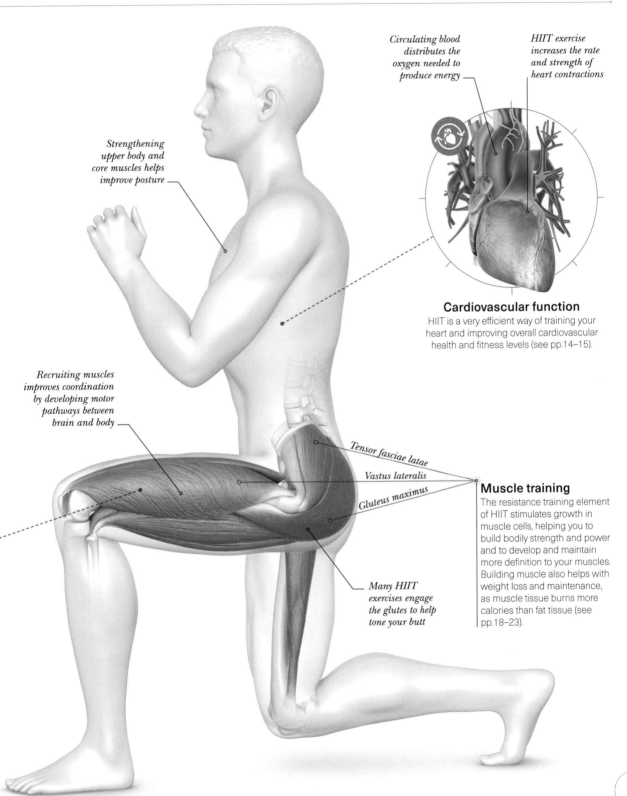

Strengthening upper body and core muscles helps improve posture

Circulating blood distributes the oxygen needed to produce energy

HIIT exercise increases the rate and strength of heart contractions

Cardiovascular function

HIIT is a very efficient way of training your heart and improving overall cardiovascular health and fitness levels (see pp.14–15).

Recruiting muscles improves coordination by developing motor pathways between brain and body

Tensor fasciae latae

Vastus lateralis

Gluteus maximus

Muscle training

The resistance training element of HIIT stimulates growth in muscle cells, helping you to build bodily strength and power and to develop and maintain more definition to your muscles. Building muscle also helps with weight loss and maintenance, as muscle tissue burns more calories than fat tissue (see pp.18–23).

Many HIIT exercises engage the glutes to help tone your butt

11

POWERING YOUR HIIT WORKOUT

Your body is beyond smart. It reacts voluntarily and involuntarily. It allows you to run, jump, lift, cycle, swim, and so much more. In order to be able to do all these things efficiently the body needs energy. The body is able to take the things we eat and without our awareness convert this fuel into energy.

ENERGY CONVERSION

The body takes the fuel that we feed it—carbohydrates, proteins, and fats—and converts it to energy in a process called respiration. The rate at which the body uses food energy to sustain life and perform various activities is called the metabolic rate. The total energy conversion rate of a person at rest is called the basal metabolic rate (BMR). The BMR is a function of age, gender, total body weight, and amount of muscle mass (which burns more calories than body fat); athletes have a greater BMR due to this last factor. The energy consumption of people during various activities can be determined by measuring their oxygen use (VO2), because most of the energy released through respiration requires the presence of oxygen for the chemical reaction. The principal means through which the cells of the body access, transfer, and store energy is via the molecule adenosine triphosphate or ATP.

Accessing ATP

The body can draw on three different systems to access ATP energy: aerobic respiration, anaerobic glycolysis, and anaerobic phosphagen. These process are all connected and work together for our survival. Without aerobic metabolism, we would lack the sources of energy needed to perform sustained daily activities. Without anaerobic metabolism our ability to snap into action, such as during a fight-or-flight scenario, would be severely compromised.

ENERGY SYSTEMS

Aerobic respiration is the primary system for powering the body and entails a chemical reaction in the presence of oxygen to convert food energy, usually glucose stores, into molecules of ATP. Anaerobic respiration creates energy without oxygen, and there are two types: phosphagen and glycolysis. The phosphagen system activates stored ATP in cells for rapid, immediate use. The glycolysis system then takes over to supply short-term energy before oxygen intake is sufficient, and also kicks in if exercise intensity outstrips the ability of your cardiovascular system to supply oxygen (VO2 max; see pp. 14–17).

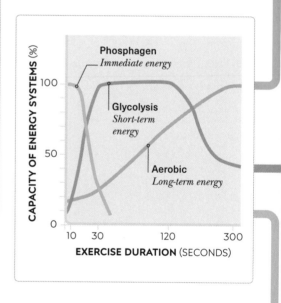

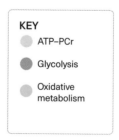

KEY

ATP–PCr

Glycolysis

Oxidative metabolism

Aerobic respiration

Occurring mainly in the mitochondria of cells (see p. 15), aerobic respiration requires oxygen to convert glucose to ATP, producing carbon dioxode and water as waste. This is the slower of the body's energy systems, but makes vastly more energy than anaerobic metabolism: some 38 molecules of ATP compared to only a maximum of 3 molecules through glycolysis. Hence why aerobic metabolism is so essential to our basic functioning, and is the primary source of energy to support low- to moderate-intensity sustained exercise. In HIIT, aerobic respiration powers the cardio moves and helps the recovery of energy after high-intensity strength moves.

ENERGY FOR ACTIVITIES
The contribution the three energy systems supply to enable a range of activities varies. The ATP–PCr system powers strength training work but other systems help replenish ATP between sets.

Anaerobic respiration: glycolysis

Anaerobic glycolysis supplies energy for high-intensity activities of moderate duration—aka a HIIT routine!—when the heart is pumping blood as fast as it can, but not enough and not in time to satisfy the oxygen needs of muscles. Occurring in the cytoplasm of cells without the presence of oxygen, glycolysis converts glucose, through a process that involves fermentation, to release only 2 or 3 ATP molecules and create a byproduct of lactate. If lactate accumulates in the blood, and cannot be cleared by aerobic respiration, lactic acidosis ensues with symptoms that include muscle ache, burning muscles, fatigue, rapid breathing, stomach pain, and even nausea. If you haven't yet felt those symptoms from an intense HIIT workout—you will! Thankfully, the symptoms of lactate burn are temporary and reversible. Once oxygen supply meets demand again, lactate can be metabolized back into pyruvate for reuse in aerobic respiration.

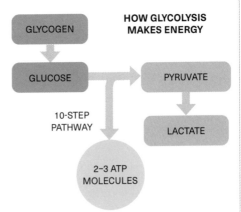

Anaerobic respiration: phosphagen

This process uses phosphocreatine (PCr) and has a very rapid rate of ATP production. The phosphocreatine is used to restore ATP after it is broken down to release energy. The total amount of PCr and ATP stored in muscles is small, so there is limited energy available for muscular contraction. It is, however, instantaneously available and is essential at the onset of activity, as well as during short-term high-intensity activities lasting about 1 to 30 seconds in duration, such as sprinting or HIIT exercises.

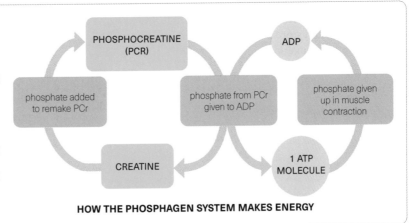

HOW THE PHOSPHAGEN SYSTEM MAKES ENERGY

IMPROVING YOUR
CARDIO FITNESS

A key indicator of fitness is how efficiently your body transports oxygen to muscles, via the cardiovascular system, in order to release the energy needed for physical activity. Embarking on a HIIT program is a great way to improve cardiovascular health.

BLOOD CIRCULATION

Aerobic respiration is the body's primary mode of generating energy. In this process, oxygen is transported through blood to cells where it is required for the chemical reactions that convert energy stores into usable energy for bodily functions—such as the muscle contractions in HIIT. The pumping action of the heart maintains the flow of blood around the body, pushing oxygen-rich blood through arteries and returning deoxygenated blood with CO_2 waste back through veins to be expelled via the lungs.

Training adaptations

HIIT workouts improve cardiovascular efficiency in several ways: by training the heart to work at a faster rate and pump more with each beat; by increasing the overall volume of blood and the quantity of oxygen-carrying hemoglobin; and by increasing the density and improving the function of capillaries around muscles.

KEY
● arteries ● veins

CARDIOVASCULAR SYSTEM

Capillaries
The delivery to muscle tissue of oxygen and nutrients in blood, and the removal of waste products such as carbon dioxide, takes place via small blood vessels known as capillaries.

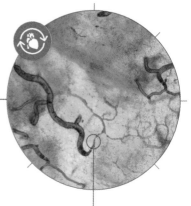

Training can increase the blood volume capacity of the heart

Red blood cells deliver oxygen

Blood
Blood delivers oxygen from the lungs and nutrients from the digestive system to facilitate the release of energy in cells, while deoxygenated blood removes the carbon dioxide that we then breathe out.

Scoring your cardiovascular capacity

A popular way of measuring aerobic fitness is to work out your "VO₂ max" score. This denotes the maximum (max) volume (V) of oxygen (O_2) the body can consume—and therefore the quantity of oxygen available in the muscles for aerobic cell respiration—during all-out physical effort. Assessing your VO₂ max can help you decide the level of HIIT routines to start with. As you progress, retesting your score provides a benchmark for tracking progress.

Take the Cooper test

Developed by Dr. Ken Cooper in 1968, the Cooper test is a simple way to measure your VO₂ max. To complete it, run as far as you can in 12 minutes and use the total distance you have run to calculate your VO₂ score, following the mathematical formula below (using either the miles or km, as appropriate).

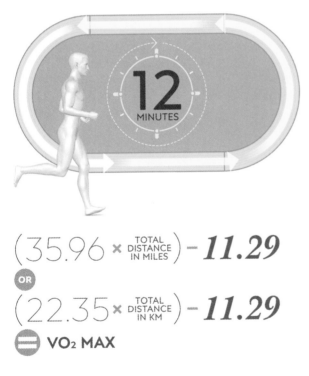

$$\left(35.96 \times \text{TOTAL DISTANCE IN MILES}\right) - 11.29$$

OR

$$\left(22.35 \times \text{TOTAL DISTANCE IN KM}\right) - 11.29$$

= VO₂ MAX

PERFORMING THE TEST
For an accurate result, run on as flat and even a surface as possible—an athletics track is ideal. Set your timer to count down from 12 minutes, run as far as possible, and record the total distance.

MITOCHONDRIAL
FUNCTION

Mitochondria are organelles within a cell that regulate metabolic activity and generate chemical energy; they are present in muscle fibers and crucial to the performance of physical activity. Multiple studies show that mitochondrial function improves in response to endurance exercise, and some studies indicate high-intensity exercise may provide a greater stimulus than moderate. In short, undertaking a HIIT program improves your capacity to generate energy at a cellular level.

Anti-aging effect

Mitochondrial function is known to decrease with age and is associated with diabetes, cardiovascular disease, and Alzheimer's. Stimulating the synthesis of mitochondria through exercise may therefore help support health in old age.

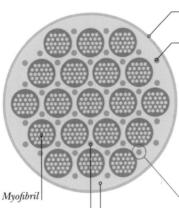

Sarcolemma

Subsarcolemmal mitochondrion

Muscle fiber
Subsarcolemmal mitochondria sit below the sarcolemma, the plasma membrane that surrounds a fiber. Intermyofibrillar mitochondria sit between the rodlike myofibrils that house the contracting filaments of muscle.

Myofibril

Intermyofibrillar mitochondrion | *Sarcoplasm*

Generating energy
The first stage of energy release takes place in the sarcoplasm, where glucose is converted to pyruvic acid. This acid then moves to the mitochondria where, via chemical reactions in the presence of oxygen, it is turned into ATP (see p.12).

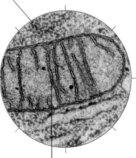

"Cristae" folds increase the surface area for the aerobic synthesis of ATP

THE HIIT "AFTERBURN"

Even though HIIT workouts are short, they are much more effective at burning calories than, say, a long, steady run. The main reason for this is that the combination of bursts of maximal effort with frequent short rests causes an extended recovery period of what scientists call Excess Postexercise Oxygen Consumption (EPOC), a.k.a. "the afterburn effect."

WHAT IS **EPOC?**

As we have seen, to power muscle contractions needed to perform the movements in HIIT exercises, the body converts stored glucose to molecules of ATP (see pp.12–15). The primary mode of conversion requires the presence of oxygen in the chemical reaction, which is supplied via the cardiovascular system. Yet even after exercise, your body has an elevated requirement for energy—and therefore oxygen—in order to fuel various processes that replenish lost glycogen stores and generally return the body to homeostatic balance. This recovery period is when Excess Postexercise Oxygen Consumption occurs to facilitate the body's raised metabolism while it adjusts down to a state of rest. As the graphs opposite demonstrate, the duration of EPOC after a short HIIT workout is far longer than after an extended session of even moderate steady-state aerobic exercise.

Fueling the recovery

With HIIT, the extended period of raised metabolism can continue beyond EPOC for up to 24 hours after a workout. To maintain metabolic efficiency, it is important to consider what you eat and when. Skipping meals or eating at widely spaced intervals, back-loaded toward the end of the day, can cause your metabolism to slow, with a drop in blood sugar levels leading to low energy. Poor nutrition can also impede physical adaptations promoted by exercise, such as muscle growth and increased glycogen stores. I would recommend eating four to five small meals balanced with the right macronutrients (see pp.26–27) and evenly spaced through a workout day. I like to use the analogy of feeding a fire with a solid piece of wood every few hours to keep it burning brightly all day.

KEY

- Oxygen deficit
- Exercise VO_2
- Recovery VO_2

PROCESSES DURING **EPOC**

During EPOC recovery, various physiological processes kick in to bring the body back to a state of rest, in which the basal metabolic rate is sufficient for energy needs. Higher oxygen levels and raised metabolism are required to reduce your heart rate, to bring down your breathing rate, and to bring your core body temperature back down to a normal 99°F (37°C). In addition, EPOC is required to restock depleted energy stores and start the work of physiological adaptation to the stimulus of exercise, including muscle growth and improved respiratory efficiency.

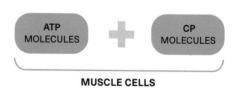

ATP MOLECULES ➕ **CP MOLECULES**

MUSCLE CELLS

ENERGY STORES IN MUSCLE CELLS
Muscle cells contain small stores of ATP and CP molecules, which supply the chemical energy for short bursts of physical effort, and these are replenished during EPOC.

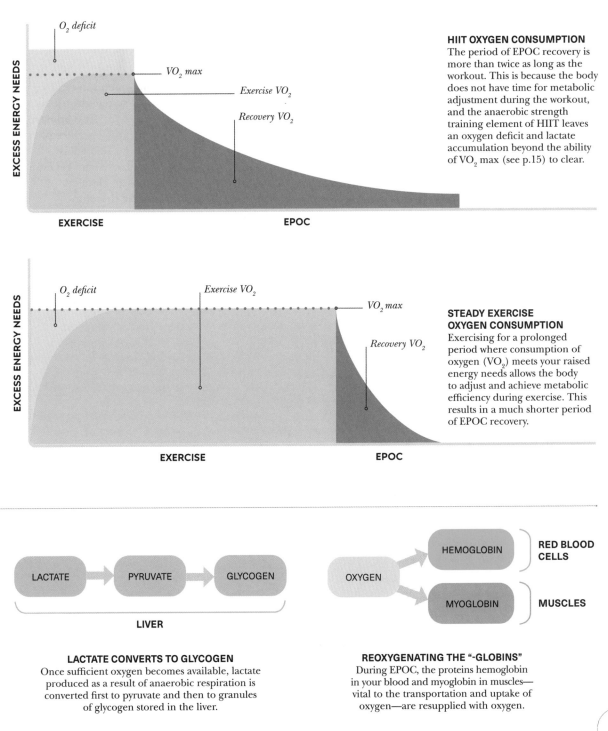

EXCESS ENERGY NEEDS

O_2 deficit

VO_2 max

Exercise VO_2

Recovery VO_2

EXERCISE

EPOC

HIIT OXYGEN CONSUMPTION
The period of EPOC recovery is more than twice as long as the workout. This is because the body does not have time for metabolic adjustment during the workout, and the anaerobic strength training element of HIIT leaves an oxygen deficit and lactate accumulation beyond the ability of VO_2 max (see p.15) to clear.

EXCESS ENERGY NEEDS

O_2 deficit

Exercise VO_2

VO_2 max

Recovery VO_2

EXERCISE

EPOC

STEADY EXERCISE OXYGEN CONSUMPTION
Exercising for a prolonged period where consumption of oxygen (VO_2) meets your raised energy needs allows the body to adjust and achieve metabolic efficiency during exercise. This results in a much shorter period of EPOC recovery.

LACTATE → PYRUVATE → GLYCOGEN

LIVER

OXYGEN → HEMOGLOBIN → **RED BLOOD CELLS**

OXYGEN → MYOGLOBIN → **MUSCLES**

LACTATE CONVERTS TO GLYCOGEN
Once sufficient oxygen becomes available, lactate produced as a result of anaerobic respiration is converted first to pyruvate and then to granules of glycogen stored in the liver.

REOXYGENATING THE "-GLOBINS"
During EPOC, the proteins hemoglobin in your blood and myoglobin in muscles—vital to the transportation and uptake of oxygen—are resupplied with oxygen.

HOW MUSCLES WORK

Muscles control movement, allowing us to do everything from jumping to chewing our food. They are attached to bones by tendons—sections of connective tissue that can resist high levels of tensile force. Muscles work in antagonistic pairs, shortening and lengthening in a cyclical way.

MUSCLE CONTRACTION

While under tension, muscles can change length (known as isotonic contraction) or remain the same (known as isometric contraction). Isotonic contractions can be either concentric or eccentric. In a concentric contraction, such as a bicep curl, the muscle shortens while generating force or overcoming resistance. In an eccentric contraction, such as lowering the body in a pull-up, the muscle lengthens while generating force. Eccentric contractions can be voluntary or involuntary.

Antagonist
The biceps brachii allows the extension of the arm

Agonist
The triceps brachii drives the extension of the arm

Extension
Angle of joint increases

Synergist
The brachialis and brachioradialis muscles assist both stages of the arm curl

ECCENTRIC CONTRACTION
During eccentric contraction, the muscle is lengthening and generating force. Eccentric contraction is stretching under tension that works to "brake" or decelerate movements. Here, the biceps brachii works eccentrically to "brake" the downward movement of the dumbbell.

HOW MUSCLES
WORK TOGETHER

In an antagonistic muscle pair, one muscle contracts as the other relaxes or lengthens. The muscle that is contracting is called the agonist and the muscle that is relaxing or lengthening is called the antagonist. For example, when you perform a bicep curl, the biceps will be the agonist as it contracts to produce the movement, while the triceps will be the antagonist as it relaxes to allow the movement to occur.

 Refining movements

Muscle coactivation is a neuromuscular response. It occurs when agonist and antagonist muscles are activated simultaneously. This type of coactivation kicks in when you are new to training, as the body tries to improve joint stability and movement accuracy. Because of this, your movements may not be the most coordinated or smooth at first. With continued practice, you can train your body to "lift" move in a more coordinated way.

Agonist
The biceps brachii drives the flexion stage

Antagonist
The triceps brachii allows the flexion of the elbow

CONCENTRIC CONTRACTION
During concentric contraction, a muscle creates tension while its muscle fibers shorten. As the muscle shortens, it generates enough force to move an object or weight. Here, the biceps brachii contracts concentrically to flex the elbow and lift the dumbbell.

Flexion
Angle of joint decreases

ISOMETRIC CONTRACTION
During isometric contraction, a muscle creates tension without any change in its length. Holding positions involve such contractions. For example, you engage abdominal muscles to stabilize your core so you can focus on the target muscles of an exercise.

Synergist
The brachialis and brachioradialis muscles assist both stages of the arm curl

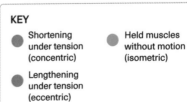

KEY

- Shortening under tension (concentric)
- Held muscles without motion (isometric)
- Lengthening under tension (eccentric)

HOW HIIT PROMOTES
MUSCLE GROWTH

HIIT workouts can build muscle, tone, and help retain lean muscle mass, as well as increasing the proportion of fast-twitch muscle fibers over slow-twitch. For muscle growth, choose mainly strength training rather than cardio routines.

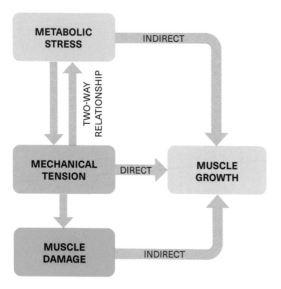

MECHANISMS OF MUSCLE GROWTH

STIMULI OF GROWTH

To build muscle you need to endure mechanical tension, muscle fatigue, and muscle damage. When you lift a heavy weight, contractile proteins in muscles generate force and apply tension to overturn the resistance. This mechanical tension is the main driver of hypertrophy (muscle growth). This tension can result in structural damage to the muscles. Mechanical damage to muscle proteins stimulates a repair response in the body. The damaged fibers in muscle proteins result in an increase in muscle size. Mechanical fatigue occurs when the muscle fibers exhaust the available supply of ATP (adenosine triphosphate; see pp. 12–13), the body's energy molecule that helps your muscles contract. They aren't able to continue fueling muscular contractions or can no longer lift the weight properly. This metabolic stress can also lead to muscle gain.

HOW MUSCLES **GET BIGGER**

Skeletal muscle protein cycles through periods of synthesis and breakdown on a daily basis. Muscle growth occurs whenever the rate of muscle protein synthesis is greater than the rate of muscle protein breakdown. Muscle hypertrophy is thought to be a collection of adaptations to different components—the myofibrils, the sarcoplasmic fluid, and the connective tissue.

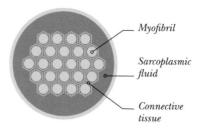

MUSCLE FIBER BEFORE GROWTH
The circle represents a muscle in cross-section with its bundle of fascicles. Within there are many myofibrils and surrounding them is sarcoplasmic fluid and a layer of connective tissue.

SATELLITE CELLS

Skeletal muscle satellite cells are considered to play a crucial role in muscle fiber maintenance, repair, and remodelling. These mononucleated cells are "wedged" between the base membrane and plasma membrane of the muscle fiber. These act as stem cells and are responsible for the further growth and development of skeletal muscles. Satellite cells go into a state of dormancy due to a sedentary lifestyle.

The decline of muscle mass with age

If you don't use them, you lose them... or they will go dormant. Our satellite cells naturally reduce in number as we get older, but training counteracts this decline. It's important for people to activate their muscles on a regular basis once they reach the age of 30, or face the possibility of losing the ability to regenerate muscle mass as they age.

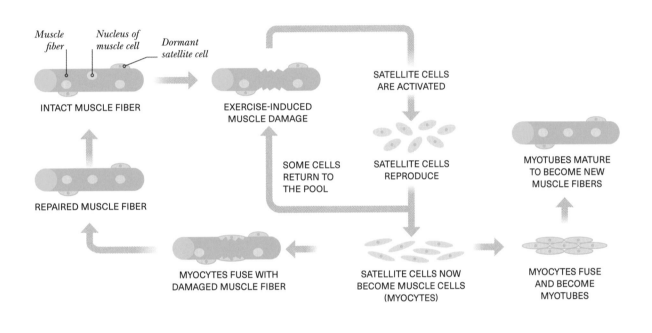

Muscle fiber · Nucleus of muscle cell · Dormant satellite cell

INTACT MUSCLE FIBER

EXERCISE-INDUCED MUSCLE DAMAGE

SATELLITE CELLS ARE ACTIVATED

SATELLITE CELLS REPRODUCE

MYOTUBES MATURE TO BECOME NEW MUSCLE FIBERS

SOME CELLS RETURN TO THE POOL

REPAIRED MUSCLE FIBER

MYOCYTES FUSE WITH DAMAGED MUSCLE FIBER

SATELLITE CELLS NOW BECOME MUSCLE CELLS (MYOCYTES)

MYOCYTES FUSE AND BECOME MYOTUBES

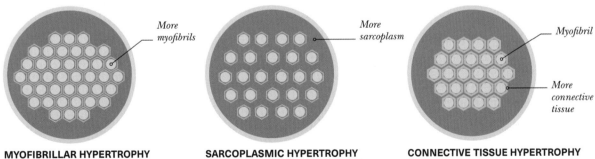

More myofibrils

More sarcoplasm

Myofibril

More connective tissue

MYOFIBRILLAR HYPERTROPHY
Myofibrillar protein makes up 60–70 percent of the protein in a muscle cell. Myofibrillar hypertrophy is the increase in number and/or size of myofibrils by the addition of sarcomeres.

SARCOPLASMIC HYPERTROPHY
A rise in the volume of the sarcoplasm (which includes mitochondria, sarcoplasmic reticulum, t-tubules, enzymes and substrates, such as glycogen) also enlarges the muscle.

CONNECTIVE TISSUE HYPERTROPHY
The extracellular matrix of the muscle is a three-dimensional scaffolding of connective tissue. Increases in its mineral and protein content lead to muscles getting bigger.

MUSCULAR
ANATOMY

There are around 600 muscles in the human body. Muscles can be categorised into three types: cardiac muscle of the heart, smooth muscle of organs, and skeletal muscle.

SKELETAL MUSCLE

Your body moves using skeletal muscles, which are attached via tendons to bones and joints, and effect movement through coordinated contractions. Studying muscles and how they move improves your mind–body connection, so you can visualize the muscles working and engage them correctly.

A zoomed-in view shows myofibrils lined up with one another

Elbow flexors
Biceps brachii
Brachialis (deep)
Brachioradialis

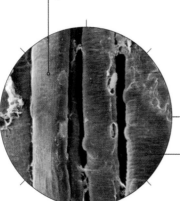

Visible stripes (striations) reflect the arrangement of muscle proteins

Skeletal muscle fibers
Like other body tissues, skeletal muscle fibersare soft and fragile. Connective tissue supports and protects the fibers, enabling them to withstand the forces of muscular contraction.

Pectorals
Pectoralis major
Pectoralis minor

Intercostal muscles

Brachialis

Abdominals
Rectus abdominis
External abdominal obliques
Internal abdominal obliques
(deep, not shown)
Transversus abdominis

Hip flexors
Iliopsoas (iliacu and psoas majo
Rectus femoris (see quadriceps)
Sartorius
Adductors (see below)

Adductors
Adductor longus
Adductor brevis
Adductor magnus
Pectineus
Gracilis

Quadriceps
Rectus femoris
Vastus medialis
Vastus lateralis
Vastus intermedius (deep, not shown)

Ankle dorsiflexors
Tibialis anterior
Extensor digitorum longus
Extensor hallucis longus

SUPERFICIAL   **DEEP**

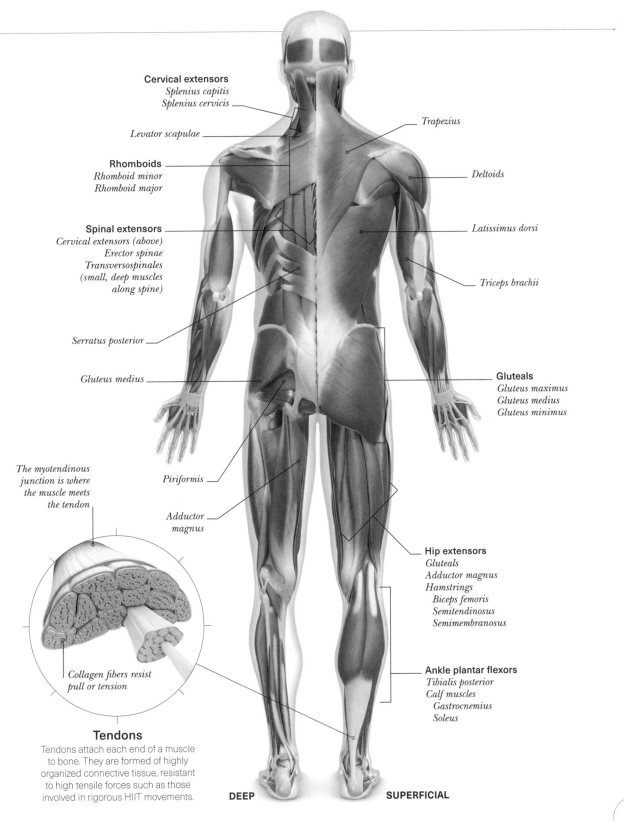

Cervical extensors
Splenius capitis
Splenius cervicis

Levator scapulae

Rhomboids
Rhomboid minor
Rhomboid major

Spinal extensors
Cervical extensors (above)
Erector spinae
Transversospinales
(small, deep muscles
along spine)

Serratus posterior

Gluteus medius

The myotendinous
junction is where
the muscle meets
the tendon

Piriformis

Adductor
magnus

Collagen fibers resist
pull or tension

Tendons
Tendons attach each end of a muscle
to bone. They are formed of highly
organized connective tissue, resistant
to high tensile forces such as those
involved in rigorous HIIT movements.

Trapezius

Deltoids

Latissimus dorsi

Triceps brachii

Gluteals
Gluteus maximus
Gluteus medius
Gluteus minimus

Hip extensors
Gluteals
Adductor magnus
Hamstrings
Biceps femoris
Semitendinosus
Semimembranosus

Ankle plantar flexors
Tibialis posterior
Calf muscles
Gastrocnemius
Soleus

DEEP

SUPERFICIAL

23

HIIT TRAINING AND **THE BRAIN**

We have only just begun to uncover how many ways exercising can positively affect the brain. Exercise pumps the brain with oxygen, releases endorphins and hormones to promote the growth of brain cells, and supports brain plasticity. In addition, it improves cognitive function, mental health, and memory, and reduces depression and stress.

IMPROVING BRAIN–BODY **CONNECTION**

Exercise has a number of positive effects on how we think and feel. Increased blood flow means that the brain is exposed to more oxygen and energy than it had prior to exercise. On an emotional level, exercise boosts our mood by sending hormones to the brain that signal happiness. According to the Harvard Medical School Journal, exercise also induces the release of beneficial proteins in the brain. These nourishing proteins (neurotrophins) maintain healthy brain cells and promote the growth of new ones.

Reduces stress: exercise is known to relieve long-term stress, and endorphins after workout cause an instant high.

Improves sleep: exercise helps to improve sleep quality, increasing the amount of rejuvenating "slow wave" deep sleep. Better sleep improves creativity and boosts brain function.

Helps protect against dementia: an increase in neurotrophins lessens brain tissue damage linked to dementia.

Improves cognitive abilities: research suggests the increase in blood flow to the brain caused from exercise can raise levels of neurotrophins, which aid the brain's ability to adapt and regenerate, improving rational thinking, intellectual performance, and memory.

Increases brain volume: research has shown that exercising enlarges the hippocampus, the area of the brain associated with memory and learning.

HIIT TRAINING **BRAIN GAINS**

When we exercise, we experience an increase in oxygen levels and angiogenesis (blood vessel growth) in the brain. In particular, this occurs in areas of the brain responsible for rational thinking and other intellectual, physical, and social abilities. Exercise also lowers levels of the stress hormones cortisol, allowing for an increase in the number of neurotransmitters like serotonin and norepinephrine.

Neurogenesis

Scientists once believed we had only a fixed amount—around 86 bilion—of neurons (nerve cells) in the brain. Research now proves that neurogenesis (the creation of new neurons) can occur and in areas such as the hippocampus, important for learning and memory. What's more, exercise has the power to stimulate levels of neurotrophins that help promote neurogeneis and neuroplasticity (see below).

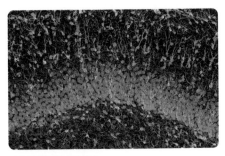

NEW BRAIN CELLS
Neuron cells are colored pink in this microscopic-level image of the brain's hippocampus. HIIT workouts promote the creation of new neurons.

Mind–body connection

When training, it's good to work with focus and without directions. One way to do this is by improving the mind-to-body connection. Mind–body connection means to focus intently on the muscle you are working on. See that muscle and feel how it moves. Research shows this boosts the strength and growth of that specific muscle. This is a mindful approach to resistance training.

Neuroplasticity

Exercise has been linked to increased neuroplasticity: the brain's ability to adapt, master new skills, and store memories and information. Pathways within the brain become more permanent the more you use them. The more you train or perform a new skill the stronger the pathways in the brain become. Training your body also trains your mind!

Neurochemistry

Where one neuron meets another there is a gap called a synapse. To transmit electrical signals, your brain uses a system of chemicals called neurotransmitters, which diffuse across the synapse and initiate the signal in the next neuron. HIIT exercises boost levels of certain neurotransmitters, such as dopamine and serotonin, which is why after working out you feel happy and less stressed.

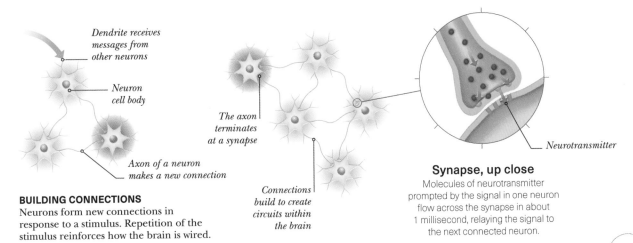

Dendrite receives messages from other neurons

Neuron cell body

Axon of a neuron makes a new connection

The axon terminates at a synapse

Connections build to create circuits within the brain

Neurotransmitter

BUILDING CONNECTIONS
Neurons form new connections in response to a stimulus. Repetition of the stimulus reinforces how the brain is wired.

Synapse, up close
Molecules of neurotransmitter prompted by the signal in one neuron flow across the synapse in about 1 millisecond, relaying the signal to the next connected neuron.

FEEDING YOUR HIIT

Even if weight loss is your primary aim in starting a HIIT program it is vital to eat enough and to eat well, to be properly energized for each workout and to reap the long-term benefits of physical transformation. Take the opportunity to embrace whole foods, variety, and a healthy balance of carbs, proteins, and fat.

THE ELEMENTS OF NUTRITION

Macronutrients are the three broad categories of nutrition: carbohydrates, proteins, and fats. Carbohydrates supply saccharides (sugars) of differing complexity, which are converted by the body to glucose and stored as glycogen, our main energy source. Protein is made up of amino acids, which are used by the body to build and repair tissues, including organs and muscle, and maintain the functioning of bodily processes. Fats are an important energy source and crucial to hormone production.

Micronutrients

Although consumed in tiny amounts, vitamins and minerals are vital to the functioning of almost all processes in the body, from immunity to cell renewal and energy production. They are best absorbed and utilized as part of whole foods rather than in supplements.

Getting the balance right

Carbohydrates should constitute the bulk of our food intake, but that doesn't mean nothing but donuts, fries, and sugary drinks! Think "whole foods" that take longer to break down for slow-release energy, and bring fiber and micronutrients with them: whole grains, vegetables, fruit (though not to excess due to high sugar), and herbs. Protein should make up around 20 percent of daily intake, whether from plant sources such as pulses, nuts, and soy, or meat, fish, and dairy. Don't avoid fat but prefer mono- and polyunsaturated.

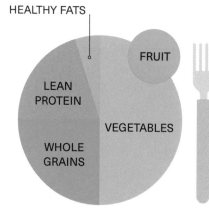

HEALTHY FATS · FRUIT · LEAN PROTEIN · VEGETABLES · WHOLE GRAINS

HEALTHY PLATE
Left is a representation of the broad proportions and mix of nutrition we should be aiming for at each main meal throughout the day.

MEASURING PORTIONS
Follow the guidance below and use your hands as a ready reckoner for the maximum portion sizes to aim for at main meals.

VEGETABLES
(TWO CUPPED HANDS)

WHOLE GRAINS
(FISTFUL)

FRUIT
(FISTFUL)

PROTEIN
(PALM SIZE)

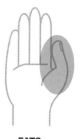

FATS
(½–1 FL OZ/1–2 TBSP)

PRE- AND POST-WORKOUT NUTRITION

On workout days especially, it is important to keep your body supplied with regular and consistent supplies of nutrition, to fuel the raised metabolism, maximize post-workout "afterburn" (see pp. 16–17), and promote effective recovery that locks in physical adaptations, such as increased muscle mass and greater glycogen stores. Everyone is different, but I would recommend avoiding eating shortly before or during training, and a snack to boost recovery shortly after working out.

WHAT TO EAT WHEN
The debate still rages about whether it is necessary to consume protein shortly after a workout to avoid excess muscle breakdown. There is, though, evidence that such intake helps to support muscle synthesis, and is recommended if you have trained in a fasted state.

EATING BEFORE

A meal or snack before training can help replenish energy stocks and leave you primed for muscle recovery. However, a full meal should be consumed 2 or 3 hours before working out, and a snack no sooner than 1 hour, so that digestive processes have worked through and don't clash with your exercising.

FASTED WORKOUT

Some people find that working out in a fasted state— for example in the morning after the natural fasting period of sleep— helps to burn more fat as the body has limited glycogen stores to draw on and turns to fat stores. This is a personal preference (I prefer it), but you may find that you *need* to eat before a workout.

EATING AFTER

You need carbohydrates after a workout to replenish energy stores, and protein to support muscle adaptation. How soon after is up to you, but don't leave it too long, and don't let a post-workout energy "bonk" be an excuse for eating refined, high-sugar snacks. Go for complex carbs, quality protein, and a healthy balance.

BEFORE TRAINING	TRAINING	AFTER TRAINING
3 HOURS 2 1	30MINS	1 2 3

Avoid eating any later than 1 hour before a workout

Some swear by pre-workout shots of apple vinegar or coffee to boost performance

Drink a protein shake straight after training

Balanced meal around 1 to 2 hours after a workout

YOUR FLUID BALANCE

Up to 60 percent of the body is water, and hydration has numerous crucial functions that can affect athletic performance. Water regulates body temperature through sweating, transports nutrients, removes the waste products of metabolism, and maintains blood flow and volume so that muscles are supplied with oxygen-rich blood for aerobic respiration (see pp. 12–15). You should therefore take care to remain well-hydrated. But do guard against overhydration, especially after exercise when you've sweated a great deal, as this can lead to a dangerous depletion in sodium.

HOW MUCH WATER TO DRINK A DAY
Current advice recommends drinking ½fl oz/lb (30–40ml/kg) of water to bodyweight, but do also adjust on a daily basis according to how much you tend to sweat, your activity level, and environmental factors.

110 LB	154 LB	220 LB
50–68 FL OZ	71–95 FL OZ	101–135 FL OZ
6–8 GLASSES A DAY	8–11 GLASSES A DAY	12–16 GLASSES A DAY

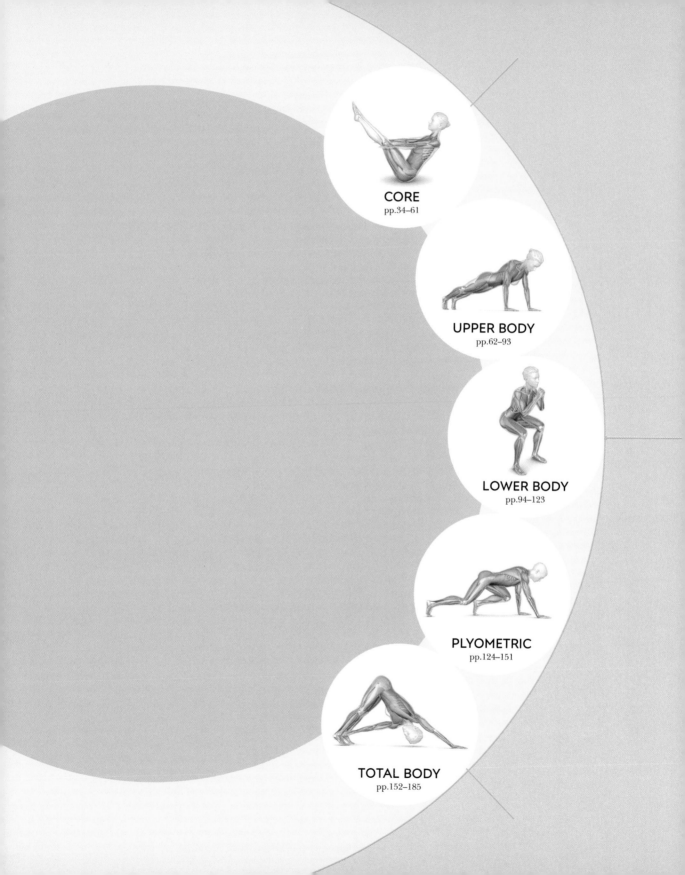

CORE
pp.34–61

UPPER BODY
pp.62–93

LOWER BODY
pp.94–123

PLYOMETRIC
pp.124–151

TOTAL BODY
pp.152–185

HIIT
EXERCISES

This exercise section features 95 exercises, comprising 46 main exercises plus 49 variations that add a challenge or modification. When performing, alternate periods between 30–60 seconds of intense exercise with 30–60 seconds of recovery. Changing the intensity and/or duration of exercise and recovery allows for an infinite number of workouts. The workouts are targeted at various muscle groups, so you can choose which to concentrate on.

INTRODUCTION TO THE EXERCISES

Performing the exercises in this section will help build both cardiovascular and muscular strength and endurance. Each exercise targets specific muscle groups that are identified in the illustrations, along with clear descriptions for how to execute with proper form and breathing techniques. Do follow the instructions carefully so that you can perform moves safely.

MAINS AND VARIATIONS

The exercises that follow are organized by area of the body, and then plyometric—for speed and power—and total body moves for agility and cardio. Within that are "main" exercises and "variations" to those mains. Each main exercise has been selected to target a specific muscle group or groups. The variations offer a way to either modify the main, to make it less challenging or to target slightly different muscles, or to create a progression to the main.

! Common mistakes

On all of the exercise pages is a common mistakes box. This box is here to mention the most common errors made when performing the selected exercises. It's important to start at your current level of skill and fitness, with low weights and duration times until you have perfected the form. Never sacrifice form by increasing the weight load.

CORRECT EXERCISE EXECUTION
Proper form is very important in the execution of each exercise. Using proper form helps to put the tension on the intended muscle to tone and increase muscle strength. In addition, this also helps avoid injury.

Brain and nervous system
The mind-to-body connection helps target the intended muscles and improve coordination.

Cardiovascular system
This carries blood enriched with oxygen to fuel the body and energize the muscles.

Respiratory system
Proper breathing helps to increase available oxygen; use the rhythm of your breath to aid form.

Muscular system
Performing these exercises correctly helps put the tension and stress on designated muscles.

Skeletal system
Muscles attach to bones and contract and relax to pull on them, causing movement. Performing exercises with proper form places stress in the right areas, helping to prevent injuries.

The respiratory system supplies the body with the oxygen required for most of its energy needs, and removes the waste carbon dioxide created from energy conversion (see pp. 12–17). Breathing also helps provide a rhythm for the mind–body connection to help you maintain control, and actively engages the core muscles, especially the abdominals.

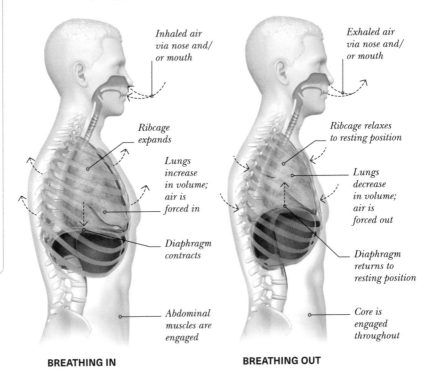

Inhaled air via nose and/ or mouth

Ribcage expands

Lungs increase in volume; air is forced in

Diaphragm contracts

Abdominal muscles are engaged

BREATHING IN

Exhaled air via nose and/ or mouth

Ribcage relaxes to resting position

Lungs decrease in volume; air is forced out

Diaphragm returns to resting position

Core is engaged throughout

BREATHING OUT

EQUIPMENT

Most of these exercises require no equipment, perfect for the home as well as the gym. A mat makes floor movements more comfortable; a ball introduces instability to engage different muscles and make you work harder; while resistance bands and dumbbells increase the load and effort required.

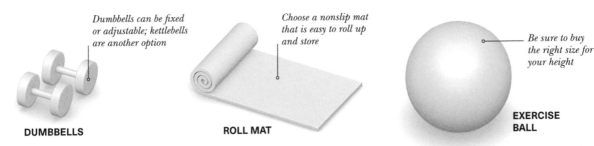

Bands are color-coded by level of resistance

RESISTANCE BANDS

Dumbbells can be fixed or adjustable; kettlebells are another option

Choose a nonslip mat that is easy to roll up and store

Be sure to buy the right size for your height

DUMBBELLS

ROLL MAT

EXERCISE BALL

TERMINOLOGY GUIDE

The body's joints facilitate a huge range of motion and the technical terms to describe these moves are explained here. Many of these terms are used to describe exercises, so I'd advise referring back to these pages.

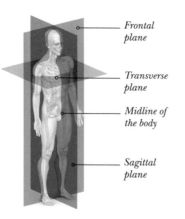

- Frontal plane
- Transverse plane
- Midline of the body
- Sagittal plane

PLANES OF MOTION

Three imaginary lines cut through the body to form planes of motion. Forward and backward motion occurs in the sagittal plane, dividing left and right halves. The frontal plane divides front and back and is the plane of side movements. The transverse plane cuts a horizontal divide, where rotational motion occurs.

Spine

As well as providing structural support for the upper body, the spine helps to transfer loads between the upper and lower body. It can extend, flex, rotate, and flex to the side, as well as combinations of these.

Neutral spine

EXTENSION
Bending at the waist to move the torso backward.

FLEXION
Bending at the waist to move the torso forward.

ROTATION
Turning the trunk to the right or left on the midline.

SIDE FLEXION
Bending the trunk to the right or left from the midline.

Elbow

The elbow is involved in any exercise using hand-held resistance as well as specific arm movements.

EXTENSION
Straightening the arm, increasing the joint angle.

FLEXION
Bending the arm, decreasing the joint angle.

Wrist

The wrist should remain neutral (in straight alignment with the forearm) unless otherwise directed.

SUPINATION
Rotating the forearm so its palm faces up.

PRONATION
Rotating the forearm so its palm faces down.

Hip

The hip joint is capable of a wide range of motion in multiple planes, all involving a straight leg, as shown here.

ADDUCTION
Moving the thigh inward toward the midline.

ABDUCTION
Moving the thigh away from the midline.

EXTERNAL ROTATION
Rotating the thigh outward.

INTERNAL ROTATION
Rotating the thigh inward.

EXTENSION
Extending the thigh backward, straightening the body at the hip.

FLEXION
Moving the thigh forward, bending the body at the hip.

Shoulder

This complex joint has a wide range of movements in multiple planes. It can move the arm forward and backward, and up and down at the side, as well as rotating at the shoulder joint itself.

FLEXION
Moving the arm forward at the shoulder.

EXTENSION
Moving the arm backward at the shoulder.

ADDUCTION
Moving the arm toward the body.

ABDUCTION
Moving the arm away from the body.

EXTERNAL ROTATION
Rotating the arm outward at the shoulder.

INTERNAL ROTATION
Rotating the arm inward at the shoulder.

Knee

The knee has to be able to sustain loads of up to 10 times the body's weight. Its main actions are flexing and extending.

FLEXION
Bending at the knee, which decreases the joint angle.

EXTENSION
Straightening at the knee, which increases the joint angle.

Ankle

In HIIT exercise, the important movements of this joint involve its dorsiflexion and plantarflexion.

DORSIFLEXION
Bending at the ankle so that the toes point upward.

PLANTARFLEXION
Bending at the ankle so that the toes point downward.

ANTERIOR VIEW

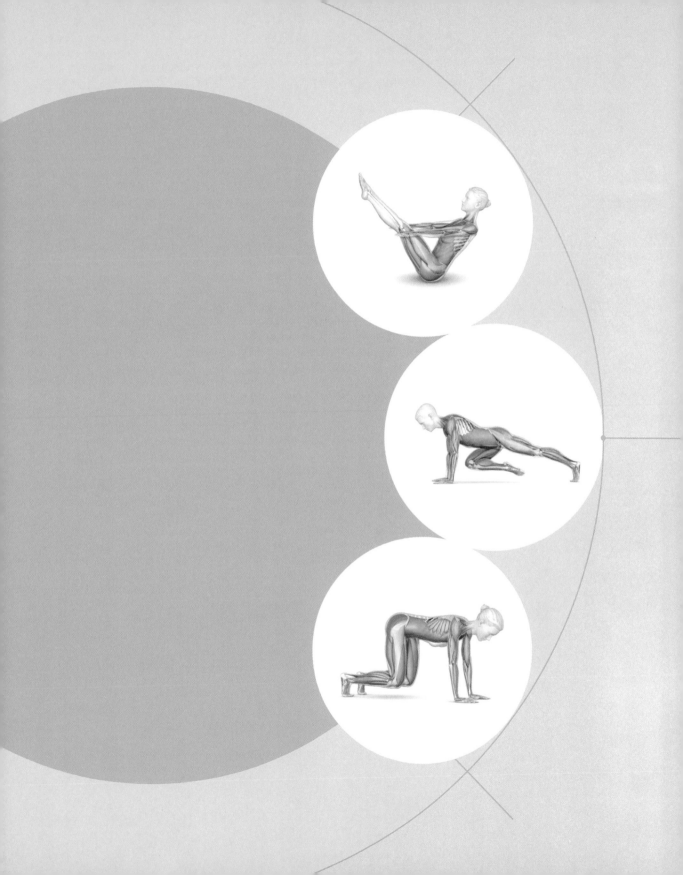

CORE
EXERCISES

Exercises in this section focus effort on abdominal muscles: the transverse abdominis, rectus abdominis, and the internal and external oblique muscles. As well as clear directions on how to execute each move to maximize effectiveness and reduce injury risk, many of the exercises include variations and modifications.

HIGH **PLANK** TO LOW **PLANK**

This exercise strengthens the abdominal muscles, the back, the glutes, the quads, and the chest. Most valuably, because of the up-and-down movement, the low plank to high plank also isolates the triceps. Performing a plank with this type of transition forces the core to engage in way that braces the back.

THE **BIG** PICTURE

All you need for this exercise is a mat. Throughout this whole workout, make sure the core is engaged—think belly button to spine, and keep your head and neck in alignment as you move up and down. Start by performing the exercise in 30-second increments with a 30-second rest. Slowly increase the time to 45 seconds, then eventually 60 seconds.

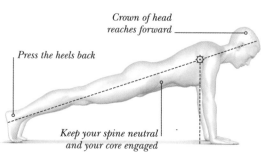

Crown of head reaches forward

Press the heels back

Keep your spine neutral and your core engaged

PREPARATION STAGE
Start in a high plank position, with arms shoulder-width apart; feet hip-width apart; and head, neck, and spine in alignment. Your fingers are flat on the floor, and your toes are bent and resting on the floor. Tuck your hip, flattening the low back curve.

Legs
The **hamstrings** help by working against gravity to keep your body in alignment. When you engage them, they help your body stay in a straight line. The **glutes**, **adductors**, and **abductors** remain tight and engaged. Squeeze the inner thighs together along with the glutes, keeping the pelvis tucked under.

KEY

- •-- *Joints*
- o— *Muscles*
- ● Shortening with tension
- ● Lengthening with tension
- ● Lengthening without tension
- ● Held muscles without motion

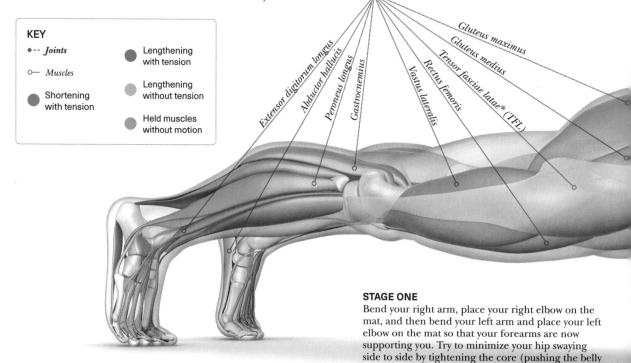

Extensor digitorum longus
Abductor hallucis
Peroneus longus
Gastrocnemius
Vastus lateralis
Rectus femoris
Tensor fasciae latae* (TFL)
Gluteus medius
Gluteus maximus

STAGE ONE
Bend your right arm, place your right elbow on the mat, and then bend your left arm and place your left elbow on the mat so that your forearms are now supporting you. Try to minimize your hip swaying side to side by tightening the core (pushing the belly button to the spine).

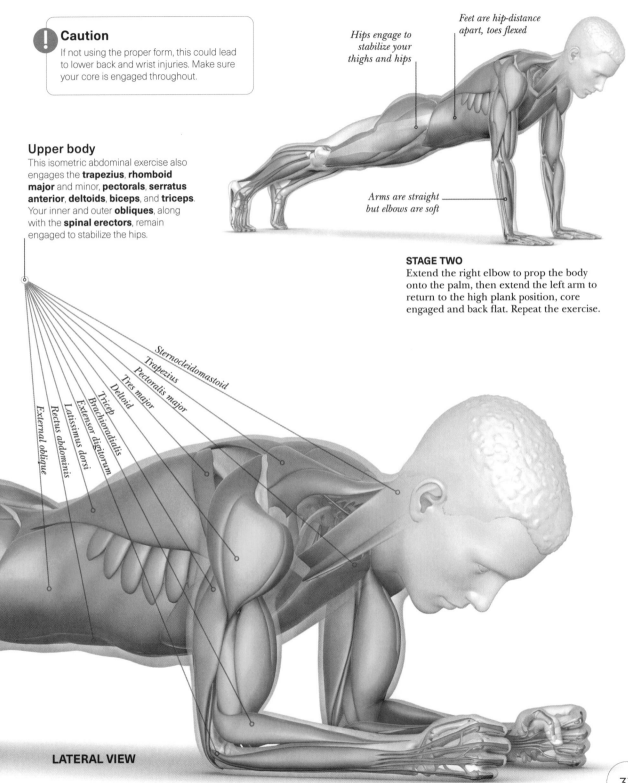

! Caution

If not using the proper form, this could lead to lower back and wrist injuries. Make sure your core is engaged throughout.

Hips engage to stabilize your thighs and hips

Feet are hip-distance apart, toes flexed

Upper body

This isometric abdominal exercise also engages the **trapezius**, **rhomboid major** and minor, **pectorals**, **serratus anterior**, **deltoids**, **biceps**, and **triceps**. Your inner and outer **obliques**, along with the **spinal erectors**, remain engaged to stabilize the hips.

Arms are straight but elbows are soft

STAGE TWO

Extend the right elbow to prop the body onto the palm, then extend the left arm to return to the high plank position, core engaged and back flat. Repeat the exercise.

Sternocleidomastoid
Trapezius
Pectoralis major
Tres major
Deltoid
Triceps
Brachioradialis
Extensor digitorum
Latissimus dorsi
Rectus abdominis
External oblique

LATERAL VIEW

» VARIATIONS

The plank variations include modified versions—low plank hold and low-impact plank, and a more advanced version, the dolphin plank. All of them target the abdominal muscles of the transverse abdominis and the rectus abdominis. The transversus abdominis must first be trained for you to build and develop your rectus abdominis or "six pack."

*Plank variations strengthen your **core**, **improve** flexibility, and **help** reduce backache.*

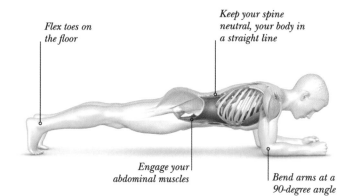

Flex toes on the floor

Keep your spine neutral, your body in a straight line

Engage your abdominal muscles

Bend arms at a 90-degree angle

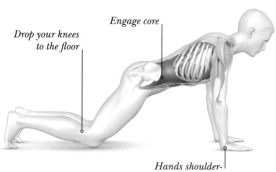

Drop your knees to the floor

Engage core

Hands shoulder-width apart

LOW PLANK HOLD

This variation involves simply holding a low plank. To avoid injuring your lower back, shoulders, neck, or hips, be sure to engage your abs, legs, and shoulders.

PREPARATORY STAGE/STAGE ONE
Get into the low plank position facing the floor, keeping the head in a neutral position, with your forearms and toes on the floor, legs stretched out, and back flat. Position your elbows directly under your shoulders and your forearms facing forward. Your head should be relaxed, your gaze toward the floor. Tuck your hips under and squeeze the glutes. Hold the low plank for 30 seconds.

LOW-IMPACT PLANK

If you have lower back issues or if you are new to HIIT workouts, the low-impact plank works best. You can still achieve the benefits of the core workout without the added spine pressure.

PREPARATORY STAGE
Start in a high plank position, hands shoulder-width apart; feet hip-width apart; and head, neck, and spine in alignment.

STAGE ONE
Maintaining a flat back and keeping your hip tucked, drop to the knees, without allowing your back to sag, and hold for 30 seconds.

KEY

● Primary target muscle

● Secondary target muscle

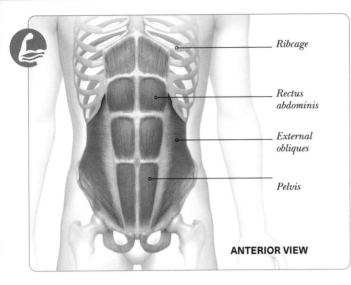

Ribcage

Rectus
abdominis

External
obliques

Pelvis

ANTERIOR VIEW

Muscles of the core

When performing a sit-up, you are using a concentric contraction. The abdominal muscles are shortening, causing the distance between the ribcage and the pelvis to decrease. However, at the top of the sit-up, when you start to lower your body to the floor, your abdominals are in an eccentric contraction, contracting under tension but not lengthening.

DOLPHIN PLANK

This is a full-body exercise. It strengthens your arms and shoulders, and you use your abs and core muscles to stabilize your torso. The hamstrings and calves get a good stretch without tension. Remember to keep your back straight, especially when you jump back.

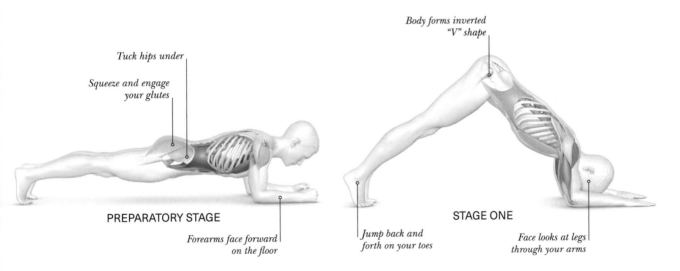

Tuck hips under

Squeeze and engage
your glutes

PREPARATORY STAGE

Forearms face forward
on the floor

Body forms inverted
"V" shape

STAGE ONE

Jump back and
forth on your toes

Face looks at legs
through your arms

PREPARATORY STAGE
Begin in the low plank position, your head neutral and facing the floor and your forearms and toes on the floor. Ensure your elbows are under your shoulders and your forearms are facing forward.

STAGE ONE
Exhale and jump your feet forward, staying on your toes while lifting your hips and bringing your body into an inverted "V" position. Maintain the position of your forearms on the floor.

STAGE TWO
Squeeze the glutes and gently jump the feet back (still on your toes) to resume the plank position on the inhale. This is the basic back-and-forth motion of the dolphin plank.

SWIM **PLANK**

The swim plank is a full-body exercise. It strengthens the abdominals, back, and shoulders. Muscles involved in the front plank include primary muscles: erector spinae, rectus abdominis (abs), and transverse abdominis.

THE **BIG PICTURE**

This exercise doesn't put as much pressure on your lower back or neck as many core exercises do. Because it includes a shift to side plank, it is a balancing exercise, so you will be building your balance and coordination. Breathe in through the nose and out through the mouth when doing the swim plank. Begin with 4 sets of 8 reps, making sure you do the same amount on each side.

Hips swing back to parallel

Core/Legs
Although the primary muscles in this movement are the **abdominals**, the **gluteus medius** and **gluteus maximus** are also activated to help stabilize your hips. In mid "swim," make sure you press the hips forward so that the spine is in neutral.

Press heels back

Squeeze your glutes and thigh muscles simultaneously

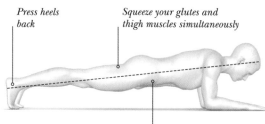

Engage the abdominals (belly button to spine)

PREPARATORY STAGE
Start in a basic low plank position, resting on the forearms, toes flexed and heels pressed back, elbows positioned directly under your shoulders and wrists in line with your elbow, also shoulder-width apart. Engage your core and keep your gaze toward the floor.

External oblique
Rectus abdominis
Tensor fasciae latae
Pectineus
Adductor longus
Rectus femoris
Vastus medialis
Vastus lateralis
Knee
Gastrocnemius
Tibialis anterior

THREE-QUARTER LATERAL VIEW

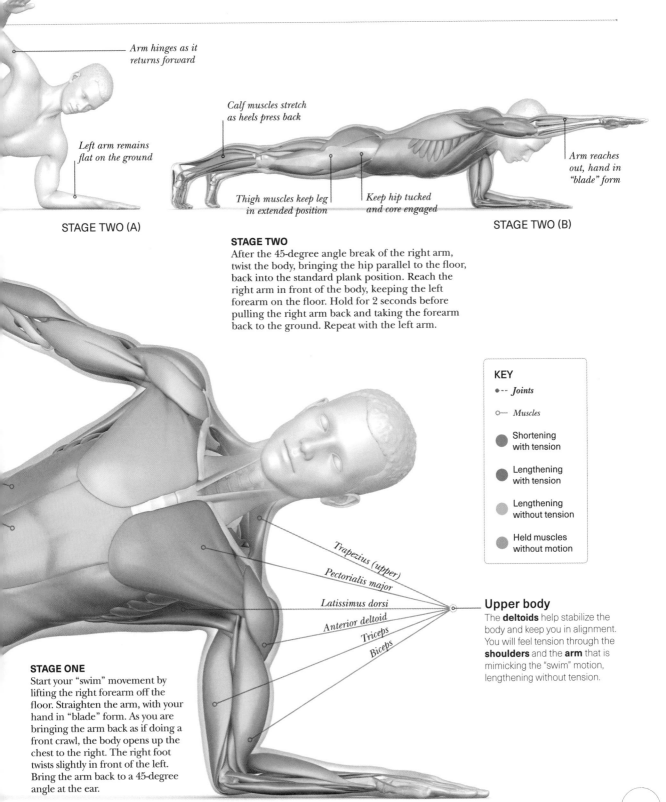

Arm hinges as it returns forward

Left arm remains flat on the ground

STAGE TWO (A)

Calf muscles stretch as heels press back

Thigh muscles keep leg in extended position

Keep hip tucked and core engaged

Arm reaches out, hand in "blade" form

STAGE TWO (B)

STAGE TWO
After the 45-degree angle break of the right arm, twist the body, bringing the hip parallel to the floor, back into the standard plank position. Reach the right arm in front of the body, keeping the left forearm on the floor. Hold for 2 seconds before pulling the right arm back and taking the forearm back to the ground. Repeat with the left arm.

KEY

●-- *Joints*

○— *Muscles*

● Shortening with tension

● Lengthening with tension

● Lengthening without tension

● Held muscles without motion

Trapezius (upper)

Pectoralis major

Latissimus dorsi

Anterior deltoid

Triceps

Biceps

Upper body
The **deltoids** help stabilize the body and keep you in alignment. You will feel tension through the **shoulders** and the **arm** that is mimicking the "swim" motion, lengthening without tension.

STAGE ONE
Start your "swim" movement by lifting the right forearm off the floor. Straighten the arm, with your hand in "blade" form. As you are bringing the arm back as if doing a front crawl, the body opens up the chest to the right. The right foot twists slightly in front of the left. Bring the arm back to a 45-degree angle at the ear.

41

MOUNTAIN **CLIMBER**

The mountain climber (also known as a "front plank with rotation") works several muscle groups and joints at the same time, strengthening your arms, back, shoulders, core, and legs. Another benefit of using multiple muscles at once is an increased heart rate, which will help you burn more calories. Your quads get a great workout, too.

Core and **arms**
The core stabilizers (**rectus abdominis**, **transverse abdominis**, and **internal** and **external obliques**) stabilize the back. The **arms** and the **shoulders** are still but under tension. The **triceps brachii** lock the elbows.

Trapezius
Deltoids
Biceps
Triceps
Spinal extensors
External oblique
Rectus abdominis

THE **BIG PICTURE**

As you perform the move, your shoulders, arms, and chest work to stabilize your upper body while your core stabilizes the rest of your body. Once in the plank position, focus on keeping your body in this straight line as you bring each knee across the body and back to plank position. If you struggle with this exercise, bring each knee straight to the chest rather than across the body. Make sure to do the same amount of reps on each side.

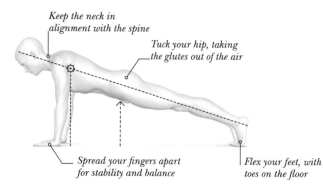

Keep the neck in alignment with the spine

Tuck your hip, taking the glutes out of the air

Spread your fingers apart for stability and balance

Flex your feet, with toes on the floor

PREPARATORY STAGE
Get into a high plank position, making sure to distribute your weight evenly between your hands and toes. Position hands shoulder-width apart, wrists under the shoulders, the neck and spine in alignment, head in a neutral position, back flat, and abs engaged. Tuck the hips so your glutes are not sticking up.

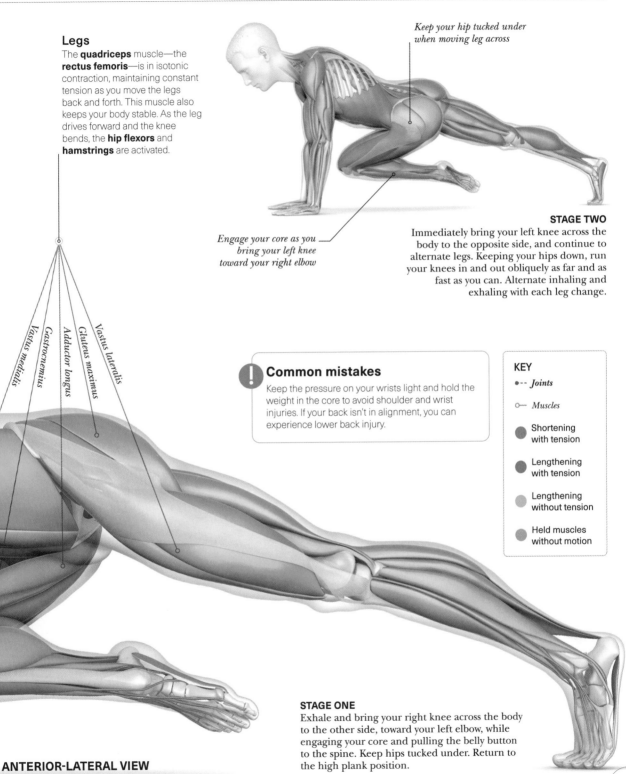

Legs

The **quadriceps** muscle—the **rectus femoris**—is in isotonic contraction, maintaining constant tension as you move the legs back and forth. This muscle also keeps your body stable. As the leg drives forward and the knee bends, the **hip flexors** and **hamstrings** are activated.

Keep your hip tucked under when moving leg across

Engage your core as you bring your left knee toward your right elbow

STAGE TWO
Immediately bring your left knee across the body to the opposite side, and continue to alternate legs. Keeping your hips down, run your knees in and out obliquely as far and as fast as you can. Alternate inhaling and exhaling with each leg change.

Vastus medialis
Gastrocnemius
Adductor longus
Gluteus maximus
Vastus lateralis

! Common mistakes

Keep the pressure on your wrists light and hold the weight in the core to avoid shoulder and wrist injuries. If your back isn't in alignment, you can experience lower back injury.

KEY

●-- *Joints*

○— *Muscles*

● Shortening with tension

● Lengthening with tension

● Lengthening without tension

● Held muscles without motion

STAGE ONE
Exhale and bring your right knee across the body to the other side, toward your left elbow, while engaging your core and pulling the belly button to the spine. Keep hips tucked under. Return to the high plank position.

ANTERIOR-LATERAL VIEW

» VARIATIONS

Mountain climber variation exercises utilize all the core muscles, including the rectus abdominis, transverse abdominis, and obliques. They also activate the muscles in the hips and back and can strengthen the lower back if done correctly.

ALTERNATING FOOT SWITCH

The explosive alternating foot switch requires fitness and coordination. Make sure you keep your back straight and flat, in alignment with the neck and head. Don't raise your hips—keep the pelvis tucked as you perform the switches.

KEY
- Primary target muscle
- Secondary target muscle

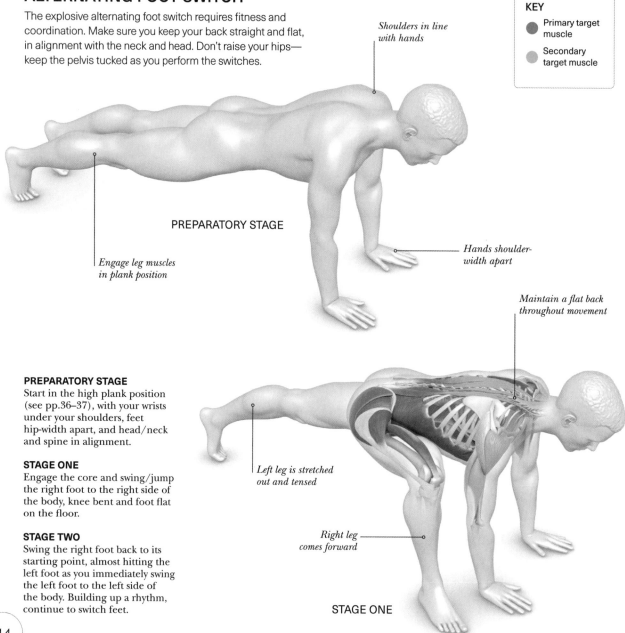

Shoulders in line with hands

PREPARATORY STAGE

Engage leg muscles in plank position

Hands shoulder-width apart

Maintain a flat back throughout movement

Left leg is stretched out and tensed

Right leg comes forward

STAGE ONE

PREPARATORY STAGE
Start in the high plank position (see pp.36–37), with your wrists under your shoulders, feet hip-width apart, and head/neck and spine in alignment.

STAGE ONE
Engage the core and swing/jump the right foot to the right side of the body, knee bent and foot flat on the floor.

STAGE TWO
Swing the right foot back to its starting point, almost hitting the left foot as you immediately swing the left foot to the left side of the body. Building up a rhythm, continue to switch feet.

66 99

*Pay attention to the **wrists and** back when you perform the mountain climber variations.*

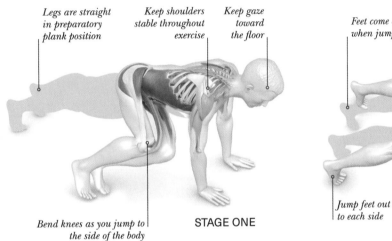

Legs are straight in preparatory plank position

Keep shoulders stable throughout exercise

Keep gaze toward the floor

Bend knees as you jump to the side of the body

STAGE ONE

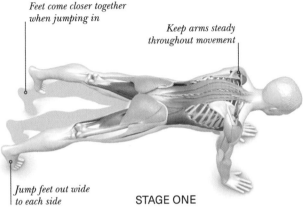

Feet come closer together when jumping in

Keep arms steady throughout movement

Jump feet out wide to each side

STAGE ONE

PLANK **SIDE-TO-SIDE JUMPS**

This side-to-side exercise is a great cardiovascular variation on the mountain climber. As well as strengthening the upper and lower abdominals and the left and right obliques, it can also improve stability, burn calories, and reduce fat.

PREPARATORY STAGE
Start in a high plank position (see pp.36–37), with your wrists under your shoulders; feet hip-width apart; and head, neck, and spine in alignment.

STAGE ONE
Bend the knees and bring them forward, jumping them to the right side of the body in a dynamic movement.

STAGE TWO
Jump the feet back to the starting position. Bend the knees and bring them forward, jumping them to the left side of the body. Repeat for 30–60 seconds, completing the same number of jumps on each side.

PLANK **JACKS**

This exercise, so called because it is like jumping jacks in plank form, helps strengthen the chest, back, arms, and shoulders. Plank jacks also strengthen effectively the muscles of the core. If you have wrist pain, you can modify the exercise by resting on your forearms.

PREPARATORY STAGE
Start in high plank position (see pp.36–37) with your arms extended and hands under your shoulders, feet together. Your body should be in a straight line from your head to your heels.

STAGE ONE
Jump both feet out wide to each side as if you were doing a horizontal jumping jack. Maintain the plank form from the preparatory stage.

STAGE TWO
Stay in the plank position as you quickly jump your feet back together, tightening your core when you jump back in. Continue to jump out and in. Keep your back flat and don't let your hips drop. Your arms should remain steady. Perform for 10–20 seconds to begin with, progressing to 60 seconds or jumping at a faster speed to add challenge.

BEAR **PLANK**

This exercise strengthens the abdominals and core muscles, reducing symptoms of lower back pain and injury risk. It also helps with balance. In addition to working the abs, the bear plank targets the muscles in your gluteus medius and maximus, psoas, quadriceps, shoulders, and arms.

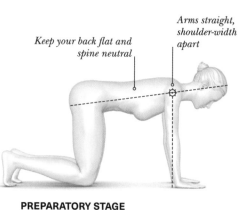

Keep your back flat and spine neutral

Arms straight, shoulder-width apart

PREPARATORY STAGE
Start on all fours (quadruped or tabletop position), making sure the back is flat. Your hands should be shoulder-width apart, wrists under the shoulders. Knees should be hip-width apart. Flex your feet, with your toes on the floor.

THE **BIG PICTURE**

When doing the bear plank, try to keep your gaze looking down toward the floor. This keeps your neck in a neutral position. Looking up toward the ceiling or out in front of you puts extra strain on your neck. Don't shift the hips back and forth—this is an isometric exercise, so stillness is important. In addition, make sure the muscles are always engaged (tight core). Increase the time holding the bear plank as you progress.

KEY

●-- *Joints*

○— *Muscles*

● Shortening with tension

● Lengthening with tension

● Lengthening without tension

● Held muscles without motion

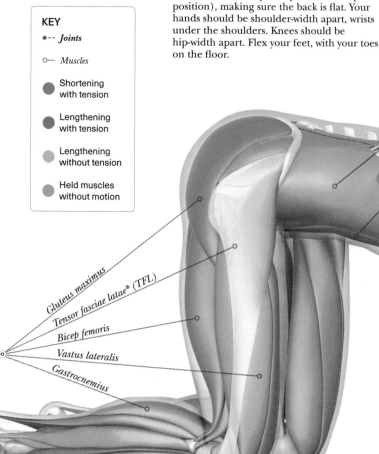

Gluteus maximus

Tensor fasciae latae* (TFL)

Bicep femoris

Vastus lateralis

Gastrocnemius

Legs
The bear plank fires the muscles in the **quadriceps**, which are held in an isometric contraction. The quadriceps muscles stabilize the body and support the weight of holding the knees above the floor. The bend in the knee engages the **hip flexors** and the **hamstrings**.

! Caution
When doing the bear plank, don't collapse your lower back, allowing it to sink in, which can strain the lower back muscles. Avoid this by keeping your core muscles engaged, your back flat, and your spine neutral.

THREE-QUARTER LATERAL VIEW

Upper body

To retain the neutral spine needed, the **abdominal muscles—transverse abdominis, rectus abdominis,** and **internal** and **external obliques—**are engaged. In addition, the **erector spinae** in the back and **psoas major** at the hips aid in this isometric hold.

External oblique
Rectus abdominis
Serratus anterior
Pectoralis major
Deltoid
Bicep
Semispinalis capitis

*The bear plank is an **isometric exercise,** meaning it engages muscles without movement.*

STAGE ONE

Engage your core (belly button to spine) to keep the back flat. Exhale and push the palms into the floor, lifting your knees 3–6in (8–15cm) off the floor. Keep your toes flexed and resting on the floor. Your hips will be level with the shoulders. Hold for 30–60 seconds, depending on your fitness level.

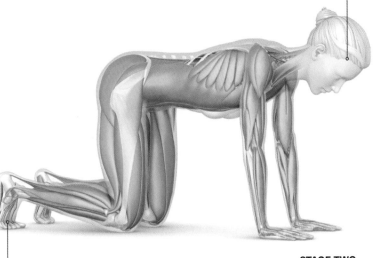

Your gaze should be on the floor

Keep your toes flexed, heels pressing back

STAGE TWO

Drop the knees back to the floor, returning to the preparatory stage.

47

» VARIATIONS

The bear plank variations give you alternatives and extra challenges beyond the basic bear plank exercise. Like the original bear plank, they work multiple muscle groups, targeting the core, glutes, hamstrings, hip flexors, and shoulders.

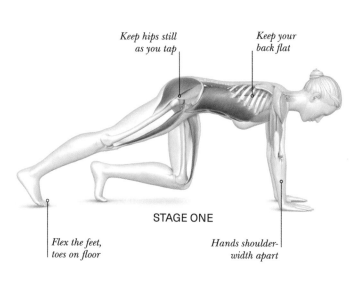

Keep hips still as you tap

Keep your back flat

STAGE ONE

Flex the feet, toes on floor

Hands shoulder-width apart

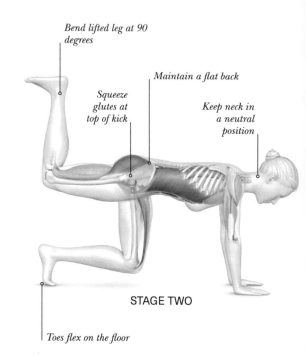

Bend lifted leg at 90 degrees

Squeeze glutes at top of kick

Maintain a flat back

Keep neck in a neutral position

STAGE TWO

Toes flex on the floor

LEG-REACH TOE TAP

The toe tap increases mobility and agility, strengthening the muscles in your gluteus medius and maximus, psoas, quadriceps, shoulders, and arms. It also works the abs.

PREPARATORY STAGE

Start on all fours in the tabletop position (see pp.46–47), making sure the back is flat. Position your hands shoulder-width apart. Make sure your wrists are under your shoulders; knees hip-width apart. Flex your feet, with your toes on the floor.

STAGE ONE

Assume the bear blank form with your knees 3–6in (8–15cm) off the floor. Reach one foot back, straightening the leg and tapping the toes on the floor.

STAGE TWO

Bring the foot back to the starting bear plank position. Repeat with the opposite foot, making sure your hips remain still and don't rock from side to side. Maintain the position of the leg that is not tapping in the bear plank position, knee lifted off the floor. Continue for 30–60 seconds.

DONKEY KICK

The donkey kick engages the glutes further while maintaining the isometric hold on the abs. One leg kicks back and up, while the knee of the other remains slightly off the floor. It's important to keep the leg that is not kicking still, as this keeps the core engaged. Alternate legs slowly so that the hip isn't shifting from left to right.

PREPARATORY STAGE

Start on all fours in the tabletop position (see pp.46–47), making sure the back is flat. The hands are shoulder-width apart, the wrists are under the shoulders, and the knees are located hip-width apart. The foot is flexed and the toes are on the floor.

STAGE ONE

Lift your knees 3–6in (8–15cm) off the floor into the bear plank position, ensuring your hips are at the same level as your shoulders. Keep your gaze on the floor and make sure your core is engaged, your back is flat, and your spine is neutral.

STAGE TWO

Bring one knee up and behind you into a donkey kick. Maintain the position of the leg that is not kicking back in the bear plank position, knee lifted off the floor. Repeat for 30–60 seconds, alternating legs in the kick.

KEY
- Primary target muscle
- Secondary target muscle

*"" As you "row" your hand or weight up, make sure the **hip** doesn't sway from side to side, but **remains as** still as possible.*

ALTERNATING ROW

The alternating row exercise works to increase mobility, agility, and focus on the back while maintaining an isometric hold and engaging the abs. You can add weights to this exercise; however, if you are a beginner, start without weight or with super-light weights. Repeat for 30–60 seconds.

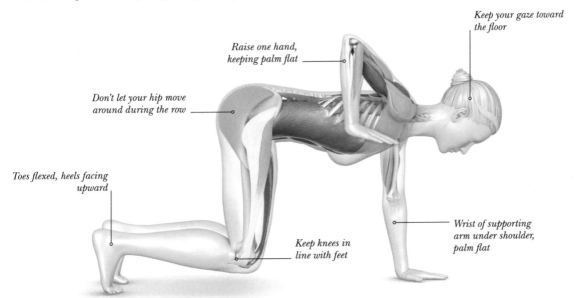

Keep your gaze toward the floor

Raise one hand, keeping palm flat

Don't let your hip move around during the row

Toes flexed, heels facing upward

Wrist of supporting arm under shoulder, palm flat

Keep knees in line with feet

PREPARATORY STAGE
Start on all fours in the tabletop position (see pp.46–47), making sure the back is flat. Position your hands shoulder-width apart, palms flat on the floor; if using weights, your hands will be on the dumbbells. The wrists are under the shoulders; knees are hip-width apart. The foot is flexed and the toes are on the floor.

STAGE ONE
Lift your knees 3–6in (8–15cm) off the floor into the bear plank position, making sure your hips are level with your shoulders. Make sure your core muscles are engaged, your back is flat, and your spine is neutral, which will keep your neck neutral. Keep your hips still, with your knees hovering above the floor.

STAGE TWO
Inhale; tighten your core; then exhale and row the right hand (or dumbbell) toward your ribcage, squeezing your shoulder blade. Keep the hips and shoulders square to the ground and don't let the hips move around. Return your hand (or dumbbell) to the floor, then repeat with the left hand.

SIT-UP

This exercise strengthens and tones the core-stabilizing abdominal muscles. It specifically works the rectus abdominis, transverse abdominis, and obliques, in addition to your hip flexors, chest, and neck. It also promotes good posture by working your lower back and gluteal muscles.

THE **BIG PICTURE**

Because the hip flexors are used when performing sit-ups, it's important to avoid having them "do all the work"—a lot of times, in order to get the torso up off the floor, the hip flexors are engaged instead of the abdominals. You need to keep the core engaged (belly button to spine) all the time. You can hold your arms on either side of your head or straight out in front of you. Start with 3 sets of 10 reps.

> **Caution**
> It's important to avoid "craning the neck," which can cause strain to the neck and back. Also, use control rather than dropping your body to the floor, which will impact your spine.

Arms raised with hands loosely by ears, not behind your head

Bend the knees

There should be no space between your back and the floor

PREPARATORY STAGE
Lie face-up on the floor. Bend your knees so your feet are flat on the floor, planted firmly. If the abdominals are on the weaker side, it's best to tuck them under a bench or brace them in another way. If working out with a partner or trainer, you can have them hold your feet down.

Upper **body** and **hips**
Sit-ups engage the **rectus abdominis**, **transverse abdominis**, and **obliques**, as well as your **hip flexors**, chest, and neck. A correct sit-up involves moving each vertebra in the spine. The **iliopsoas** and **rectus femoris** are used when bending at the hips; the **tensor fascia latae** is also engaged.

Deltoid
Serratus anterior
Pectoralis major
Rectus abdominis
External oblique
Tensor fasciae latae

STAGE ONE
Use your abdominal muscles to lift your back off of the floor. Keep your tailbone and hips static and pressed into the floor until you're fully upright. Think about lifting one vertebra at a time rather than lifting your back in one go.

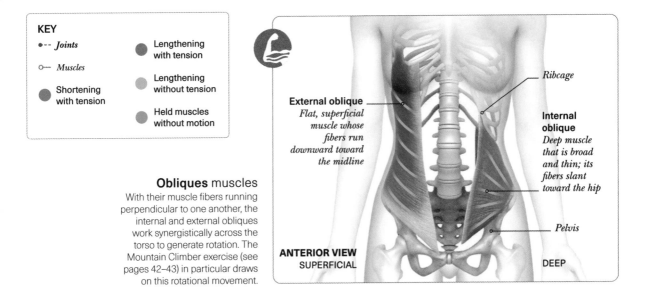

External oblique
Flat, superficial muscle whose fibers run downward toward the midline

Ribcage

Internal oblique
Deep muscle that is broad and thin; its fibers slant toward the hip

Pelvis

ANTERIOR VIEW
SUPERFICIAL

DEEP

Obliques muscles

With their muscle fibers running perpendicular to one another, the internal and external obliques work synergistically across the torso to generate rotation. The Mountain Climber exercise (see pages 42–43) in particular draws on this rotational movement.

Legs

Because your **hip flexors** are engaged, the **rectus femoris** muscle in the **quadriceps** and the **sartorius** are also activated. The **anterior tibialis**, a muscle located near the shin which acts to point and flex the foot, helps stabilize the lower body.

Rectus femoris
Bicep femoris (short)
Gastrocnemius
Soleus
Peroneus longus

Engage core all the way back to the starting position

Spine/back and neck in alignment

Plant feet on floor

LATERAL AND THREE-QUARTER ANTERIOR VIEW

STAGE TWO

With immense control, slowly lower yourself back to the starting position, uncurling one vertebra at a time, starting with your lower back. Do not let your body weight go and hit hard onto the floor.

51

CRUNCH

The crunch is one of the most widely used abdominal exercises. It is popular for working on the rectus abdominis, the "six pack" muscle that runs along the front of the torso, which is visible in those with low body fat. Building this muscle strengthens the core muscles for stability and performance.

! Caution

The number-one mistake people make when reaching fatigue is pulling on the neck. That pulling motion takes away from targeting the abs and places your neck out of alignment with the spine, possibly causing neck strain or injury. You want to originate the movement in the abs, not from the head. To keep the neck from moving, you can put your fist under your chin.

THE BIG PICTURE

Body control is important for the crunch. Control your body on the way up but especially on the way down—don't let your weight bring you to the ground with a thud. It's way more effective to keep the abdominals contracted throughout the entire movement. Always keep the spine in a neutral position—don't curve or arch it. Keep the chin up toward the thighs, core engaged, and neck and spine all in alignment with the head. Take your time and do the move slowly to start with, completing 3 sets of 10 reps. Increase the sets/reps as you progress.

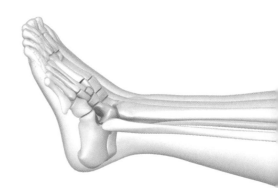

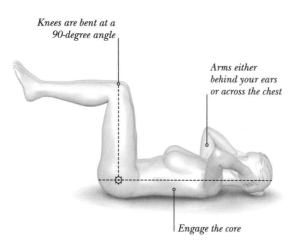

Knees are bent at a 90-degree angle

Arms either behind your ears or across the chest

Engage the core

Lower **body** and **legs**

The lower body is in the tabletop position, so to ensure that the **lower back** and **pelvic muscles** are pushed down against the floor, there should be no tension in the lower body. If you are feeling tightness in the **hip flexors**, they may need to be stretched to alleviate some tightness.

Adductor magnus

Rectus femoris

Tensor fasciae latae

PREPARATORY STAGE

Lie down on the floor on your back and bend your knees, bringing them up to a tabletop position. Place your hands behind your ears or across your chest. If you tend to pull on your neck, then across the chest or straight out in front of you is best. Pull your belly button toward your spine in preparation.

STAGE ONE

Exhale as you engage your abdominal muscles, bringing your chin toward your thighs and your shoulder blades about 1–2in (3–5cm) off the floor. You want to keep the chin parallel to the chest the entire time, at about a 45-degree angle. Hold at the top of the movement for a few seconds, breathing continuously.

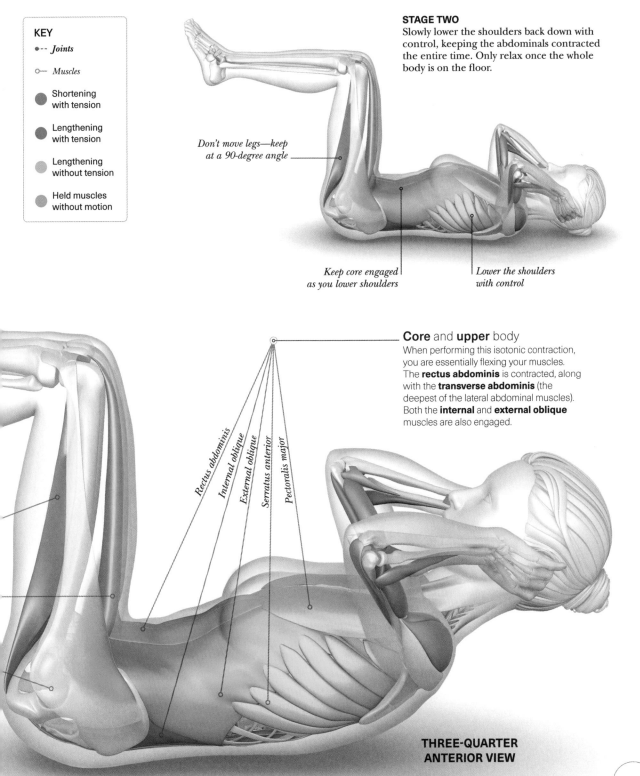

KEY

•-- *Joints*

o— *Muscles*

● Shortening with tension

● Lengthening with tension

● Lengthening without tension

● Held muscles without motion

STAGE TWO
Slowly lower the shoulders back down with control, keeping the abdominals contracted the entire time. Only relax once the whole body is on the floor.

Don't move legs—keep at a 90-degree angle

Keep core engaged as you lower shoulders

Lower the shoulders with control

Core and **upper** body
When performing this isotonic contraction, you are essentially flexing your muscles. The **rectus abdominis** is contracted, along with the **transverse abdominis** (the deepest of the lateral abdominal muscles). Both the **internal** and **external oblique** muscles are also engaged.

Rectus abdominis

Internal oblique

External oblique

Serratus anterior

Pectoralis major

THREE-QUARTER ANTERIOR VIEW

» VARIATIONS

These crunch variations will add an additional burn to the abdominals. They focus on the internal and external obliques, the transversus abdominis, and the rectus abdominis. These are perfect for any fitness level; beginners should start with 30 seconds, working up to 60 seconds.

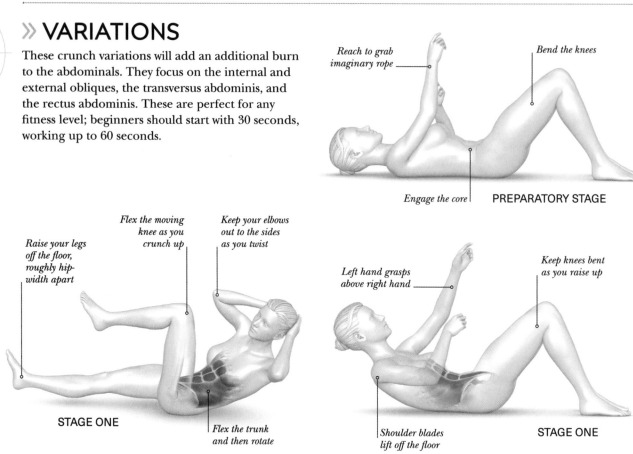

Reach to grab imaginary rope

Bend the knees

Engage the core   PREPARATORY STAGE

Flex the moving knee as you crunch up

Keep your elbows out to the sides as you twist

Raise your legs off the floor, roughly hip-width apart

STAGE ONE

Flex the trunk and then rotate

Left hand grasps above right hand

Keep knees bent as you raise up

Shoulder blades lift off the floor

STAGE ONE

BICYCLE CRUNCH

The bicycle crunch is so called because it mimics the motions of a cyclist, your legs "cycling" back and forth during the exercise. To challenge yourself, hold for 1 second at the top position, and keep both legs raised throughout the set.

PREPARATORY STAGE
Lie flat on your back with your hands lightly positioned behind your head and legs slightly flexed at the hips and knees. Lift your head off the floor slightly.

STAGE ONE
Inhale to engage the core, then exhale as you lift your left knee up and bring your opposite elbow toward it. Flex at the trunk and rotate your upper body toward your leg.

STAGE TWO
Inhale as you return to the starting position with control. Repeat, raising the opposite leg and elbow, and complete an equal number of reps on each side.

ROPE PULL

This exercise requires a lot of control, both on the way up and on the way down. The strength is coming from your abdominal muscles to pull yourself up on the "rope" and to lower yourself back down.

PREPARATORY STAGE
Lie down on your back with your knees bent, feet flat on the floor, and arms alongside your body. Pull your belly button toward your spine in preparation for the movement, engaging the core. Pretend that there is a rope dangling above your nose.

STAGE ONE
Using the right hand, reach up toward the left and grab the "rope," pulling the body off the floor. Now reach above the right hand with the left hand, angling slightly toward the right, and grab the "rope" with it, raising your body farther up into in a crunch position.

STAGE TWO
Still holding the "rope" but now descending hand after hand, slowly and with control lower your body toward the floor, focusing on your upper abs.

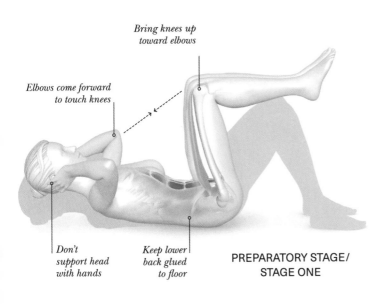

Bring knees up toward elbows

Elbows come forward to touch knees

Don't support head with hands

Keep lower back glued to floor

PREPARATORY STAGE/
STAGE ONE

DOUBLE CRUNCH

The double crunch is the joint movement of the legs and the chest. It targets a range of musculature in your core, including the rectus abdominis, rectus femoris, and obliques.

PREPARATORY STAGE
Lie on your back with your knees bent and your feet flat and hip-distance apart on the floor. Gently place your fingertips on the sides of your head.

STAGE ONE
Brace your core by engaging your abdominals. Slowly raise your knees until your thighs are just past 90 degrees with the floor. At the same time, lift your head and shoulders off the floor and raise your chest toward your knees. At the top of the rep, your forehead should be located around 6in (15cm) from your knees.

STAGE TWO
Reverse the movement to return shoulders, back, and feet to prep stage. Reset and repeat the exercise.

DOUBLE CRUNCH HOLD WITH TWIST

The double crunch hold targets the rectus abdominis and rectus femoris, while the twist works the external and internal obliques.

> **KEY**
> ● Primary target muscle
> ● Secondary target muscle

Keep hands relaxed

Flex the feet

Keep elbows stuck to midthigh in hold

Keep lower back glued to floor

STAGE ONE

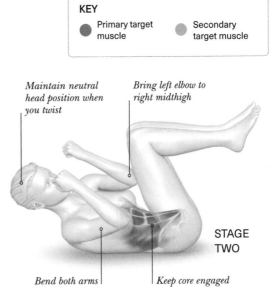

Maintain neutral head position when you twist

Bring left elbow to right midthigh

Bend both arms

Keep core engaged

STAGE TWO

PREPARATORY STAGE
Lie on your back with your knees bent at a 90-degree angle in the tabletop position, toes pointed toward the ceiling. Hold your forearms toward the face, arms bent at the elbows.

STAGE ONE
Raise your head and shoulders off the floor and bring the legs forward at the same time so your elbows are touching your midthigh. Hold this position for as long as possible, ideally up to 60 seconds.

STAGE TWO
Keeping the shoulder blades off the floor, slowly twist the body to the left and to the right, brushing the elbows toward the thighs. Left elbow touches right midthigh; right elbow touches left midthigh.

TRANSVERSE ABDOMINAL BALL CRUNCH

This exercise targets the muscles in the abdominals, including the transverse abdominis muscles that lie deep under the rectus abdominis. While the rectus muscles constitute the "six pack," for truly strong abs, you need to target both sets.

THE BIG PICTURE

You'll need an exercise ball with a diameter of at least 21½–26in (55–65cm) for this abdominal crunch. By destabilizing the surface you are lying on to perform the crunch, an exercise ball provides an activation boost, helping engage all the major abs muscles, including the obliques and smaller muscles that stabilize the spine.

Keep your head neutral; do not strain your neck

Place your hands flat on your torso

Position yourself so your glutes hang off the ball

Align your knees over your ankles

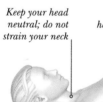

PREPARATORY STAGE
Start by sitting on top of the ball, with only the lower part of your back in contact, and with your feet shoulder-width apart, flat on the floor, and squared up to the ball. Then carefully lower the upper body so that you are in a supine position.

STAGE ONE
Inhale to engage the abs and stabilize the core. Start the crunching motion by using your abs to flex at the spine as you breathe out. Once the abs are fully flexed and you've exhaled, that is the end of the range of motion; you want to slightly raise the torso without flexing upward at the hips. To add to the challenge, hold your body at the top position for 1 second.

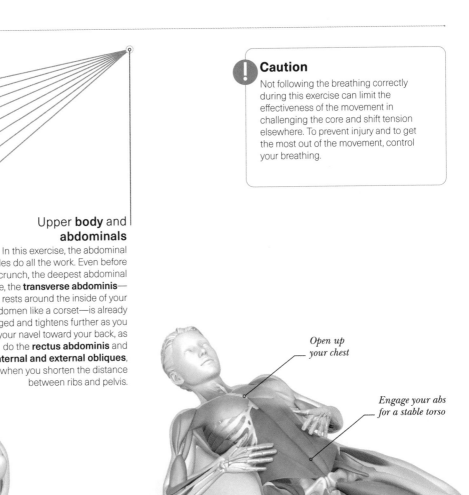

Deltoids
Pectoralis major
Brachialis
Serratus anterior
Rectus abdominis
Transversus abdominis
Internal oblique
External oblique

Caution

Not following the breathing correctly during this exercise can limit the effectiveness of the movement in challenging the core and shift tension elsewhere. To prevent injury and to get the most out of the movement, control your breathing.

Upper **body** and **abdominals**

In this exercise, the abdominal muscles do all the work. Even before you crunch, the deepest abdominal muscle, the **transverse abdominis**—which rests around the inside of your abdomen like a corset—is already engaged and tightens further as you pull your navel toward your back, as do the **rectus abdominis** and **internal and external obliques**, when you shorten the distance between ribs and pelvis.

Open up your chest

Engage your abs for a stable torso

Check your shins are perpendicular to the floor

KEY

- ●-- *Joints*
- ○— *Muscles*
- ● Shortening with tension
- ● Lengthening with tension
- ● Lengthening without tension
- ● Held muscles without motion

SUPERIOR-ANTERIOR-LATERAL VIEW

STAGE TWO

While keeping the core isometrically held, breathe in as you start to slowly lower the body back to the starting position. Repeat stages 1 and 2.

V-UP

This strength-based exercise, so-called from the shape your body makes while doing it, utilizes your body weight to target the core area. The V-up (or "hollow body") targets the abdominal muscles, toning the obliques and strengthening back muscles. At the same time, it works on your quadriceps and hamstrings and helps with balance.

THE **BIG PICTURE**

You don't need any equipment for this exercise—you just begin by lying on the floor, facing up, making sure there is no space between your lower back and the floor. It is important to maintain balance and coordination. Make sure the back isn't curved as you lift the body off the floor. Keep your back straight, using the abs and sitting bone for balance and stability. If the full version is too challenging, you can bend your knees at a 90-degree angle and bring them toward your chest, lowering them back into the straight position. To challenge yourself further, you can do the V-up on an unstable surface, such as a BOSU ball or a balance disc.

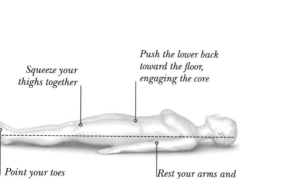

Squeeze your thighs together

Push the lower back toward the floor, engaging the core

Point your toes and rest your heels on the floor

Rest your arms and shoulder blades on the floor in a relaxed position

PREPARATORY STAGE
Start by lying face up, with your lower back glued to the floor. Make sure your legs are stretched out straight and your arms straight along either side of the body. Keep your head and spine in a neutral position.

Common mistakes
If your back isn't in alignment, you can experience lower back pains. It can also pull on the hip flexors.

STAGE ONE
In one movement, simultaneously lift your torso and legs, keeping the legs straight, and reach your arms forward. Your torso and thighs should form a "V" shape. Make sure you are pushing your back toward the floor on the way up in order to engage the core. When you lift your torso, keep your arms parallel to the floor. Don't point your fingers to your toes.

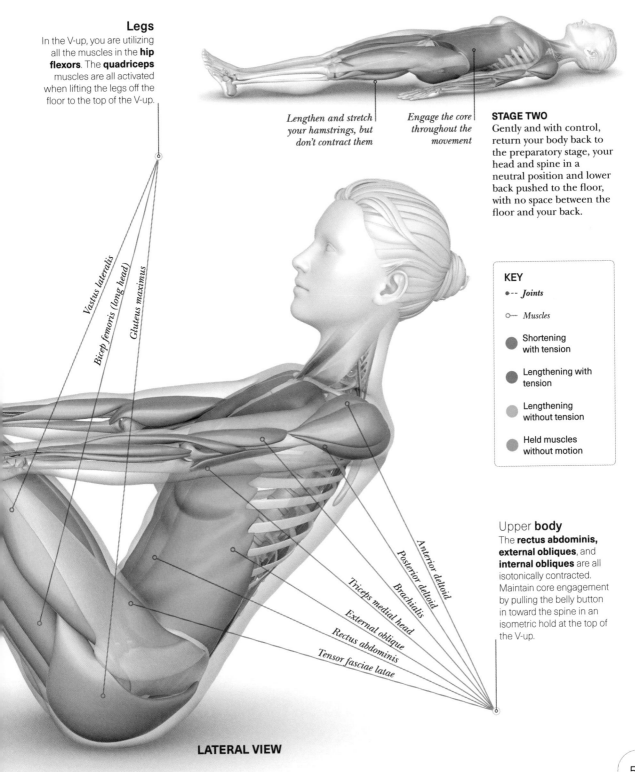

Legs

In the V-up, you are utilizing all the muscles in the **hip flexors**. The **quadriceps** muscles are all activated when lifting the legs off the floor to the top of the V-up.

Lengthen and stretch your hamstrings, but don't contract them

Engage the core throughout the movement

STAGE TWO

Gently and with control, return your body back to the preparatory stage, your head and spine in a neutral position and lower back pushed to the floor, with no space between the floor and your back.

Vastus lateralis

Bicep femoris (long head)

Gluteus maximus

KEY

- ●-- *Joints*
- ○- *Muscles*
- ● Shortening with tension
- ● Lengthening with tension
- ● Lengthening without tension
- ● Held muscles without motion

Upper **body**

The **rectus abdominis, external obliques**, and **internal obliques** are all isotonically contracted. Maintain core engagement by pulling the belly button in toward the spine in an isometric hold at the top of the V-up.

Anterior deltoid

Posterior deltoid

Brachialis

Triceps medial head

External oblique

Rectus abdominis

Tensor fasciae latae

LATERAL VIEW

⟫ VARIATIONS

The side V-up is a targeted exercise that focuses on the abdominal muscles. The primary muscles recruited include the external obliques, internal obliques, and rectus abdominis. Scissor kicks help strengthen the core, as well as other muscle groups in your torso and hips. The around-the-world V-up combines the basic V-up with the side V-ups. These variations are the perfect addition to an ab HIIT day or to a lower or upper body–focused HIIT workout.

"

*V-up variations are a **great addition to an ab HIIT workout,** and you don't need any equipment.*

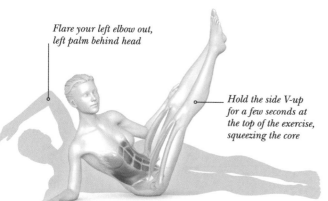

Flare your left elbow out, left palm behind head

Hold the side V-up for a few seconds at the top of the exercise, squeezing the core

STAGE ONE

The quadriceps are engaged here to aid in the raising of the leg

STAGE ONE

SIDE V-UP

This strength-based exercise utilizes your body weight to target your core area. Performing V-ups from right to left targets the abdominal muscles, specifically the corresponding obliques. It also works on your quadriceps and hamstrings, as well as helping with balance and flexibility in the hips and spine.

PREPARATORY STAGE
Start by lying on your right side, on the right hip. Your right elbow should be on the floor, under the right shoulder. Keep your hips tucked under (pelvic tilt) and legs straight, with one foot on top of the other. Bring your left arm above your head in an upside-down "V" shape.

STAGE ONE
In one movement, simultaneously lift your hips off the floor with the legs and bring your left arm alongside the legs as they lift. Your torso and thighs should form a side letter "V." Keep the back straight and use the "grounded" arm to maintain stability.

STAGE TWO
Return your legs to the starting position, lying on your side on your hip. Repeat the sequence, starting on the left side with your left elbow on the floor.

SCISSOR KICK

The scissor kick exercise works to strengthen your core, glutes, quads, and adductors. Engaging your core muscles is what allows you to "flutter" your legs up and down. The core muscles include the rectus abdominis, obliques, transverse abdominis, and hip flexors.

PREPARATORY STAGE
Lie flat on your back with your legs extended out on the floor in front of you. Place your arms by your sides, palms down. Glue your lower back to the floor. You can modify the exercise by placing your hands under your glutes below the small of your back, palms pressing into the floor.

STAGE ONE
Exhale as you lift both legs off the floor at about a 45-degree angle. With a tight core and relaxed neck, lower one leg toward the floor as you lift the other leg up.

STAGE TWO
Repeat the movement with the other leg, and continue the scissoring motion, keeping your head, neck, and spine in alignment and the lower back pressed against the floor. Minimize the "flutters" if you find this difficult.

V-UP **AROUND** THE **WORLD**

This exercise combines left and right side V-ups with a basic V-up to create an intense exercise of continual motion. You need good hip mobility and spine alignment for each around the world exercise, with the head, neck, and spine in a neutral position and the pelvic bone tucked under. The core is tight and the feet are touching. Perform the sequence of stages 1, 2, and 3, doing 2 reps at each stage.

Squeeze the quadriceps together to improve stability as you kick

Make sure the elbow of the supporting arm is under the shoulder

PREPARATORY STAGE/STAGE ONE
Lie on your right side, your right elbow on the floor, under the right shoulder. Keep your hips tucked under and legs straight. Bring your left arm above your head in an upside-down "V" shape. Lift your legs and torso up into the side V-up. Repeat, then move to stage 2.

Bring the resting arm off the floor in the middle V-up

Point the toes

Focus on the upper abdominals

STAGE TWO
Roll over to your back to perform the basic "middle" V-up, lifting up torso and legs and stretching your arms in front of you. Your torso and thighs should form a "V" shape. Keep your arms parallel to the floor when you are in this position. Repeat, then move to stage 3.

Shifting to the opposite side, support yourself on your left arm

Keep the legs straight, pointing to the left

Tighten the glutes here in addition to the abs to prevent the hips from being overworked

STAGE THREE
Roll to the left side to perform the V-up on the other side. After 2 reps, return to the right V-up and repeat the sequence.

KEY

● Primary target muscle

● Secondary target muscle

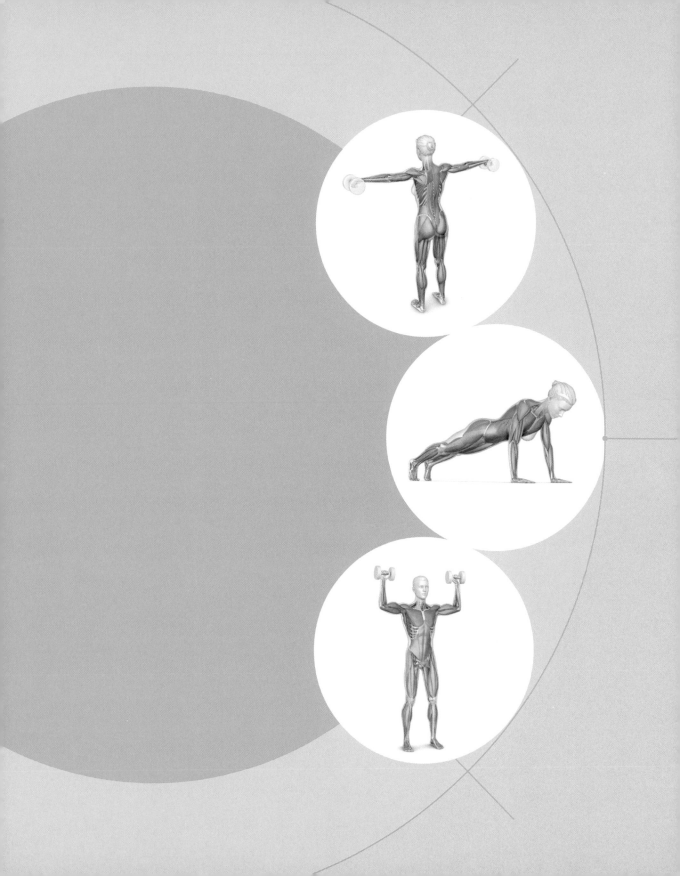

UPPER BODY
EXERCISES

Exercises in this section focus on the upper half of the body.
The moves are designed to tone and ignite the muscles in the
shoulders, biceps, triceps, back, and chest. Most of the exercises
include variations and modifications for every fitness level. The
upper body section will walk you through how to execute each
movement to maximize effectiveness and minimize risk.

PUSH-UP

This exercise strengthens the pectoral chest muscles, deltoids of the shoulders, triceps (back of your arms), the serratus anterior "wing" muscles directly under your armpit, and the abdominals. The legs act as stabilizers to prevent sagging or arching of the spinal column.

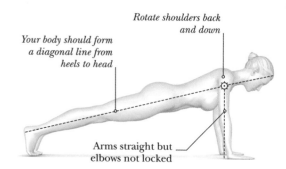

Rotate shoulders back and down

Your body should form a diagonal line from heels to head

Arms straight but elbows not locked

PREPARATORY STAGE

To begin, get into a high plank position with your pelvis tucked in, your neck neutral, and your palms slightly wider than shoulder-width apart. Make sure your shoulders are rotated back and down, too, and that the core is engaged. Flex the toes and push the heels back.

THE **BIG PICTURE**

A push-up is much more than just a chest exercise—the muscles engaged are all-encompassing. When performing a push-up, form and control are paramount. Control your body on the way down—don't allow it to just fall to the floor. Make sure the abdominals are engaged the whole time. Beginners can start with 4 sets of 5–6 reps. Look at variations (see pp.62–63) for modifications for beginners.

> **! Caution**
> You must keep the abs engaged, belly button to spine, for the duration of your push-up. This prevents the spine from caving in and putting pressure on the lower back and joints.

Peroneus longus

Gastrocnemius

Tensor fasciae latae

Vastus lateralis

Gluteus maximus

Gluteus medius

Lower **body**
The **gluteus maximus** muscle is responsible for holding the hips in place, preventing them from sagging forward and taking the spine along with them. The **rectus femoris (quadriceps** muscle) is isometrically held.

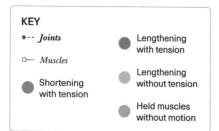

KEY

- • - - *Joints*
- ○— *Muscles*
- ● Shortening with tension
- ● Lengthening with tension
- ● Lengthening without tension
- ● Held muscles without motion

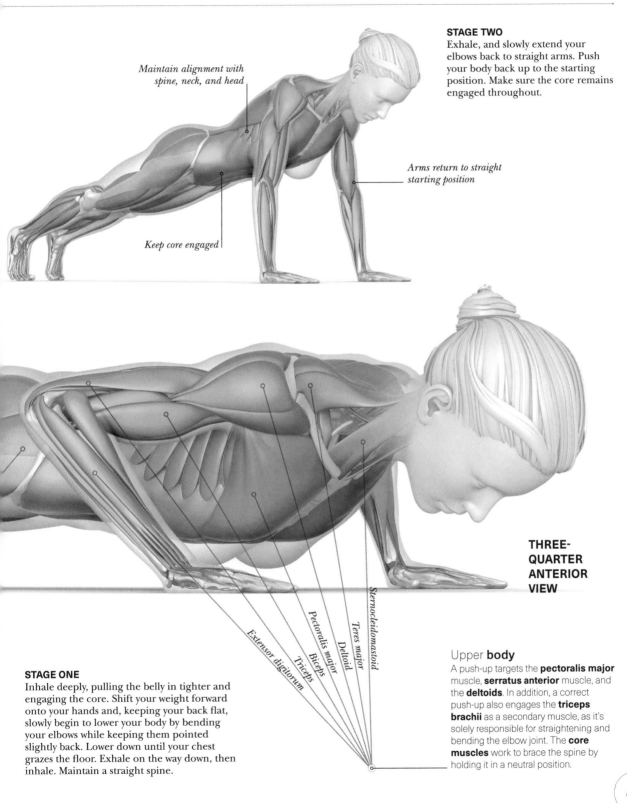

STAGE TWO
Exhale, and slowly extend your elbows back to straight arms. Push your body back up to the starting position. Make sure the core remains engaged throughout.

Maintain alignment with spine, neck, and head

Arms return to straight starting position

Keep core engaged

THREE-QUARTER ANTERIOR VIEW

Sternocleidomastoid

Teres major

Deltoid

Pectoralis major

Biceps

Triceps

Extensor digitorum

STAGE ONE
Inhale deeply, pulling the belly in tighter and engaging the core. Shift your weight forward onto your hands and, keeping your back flat, slowly begin to lower your body by bending your elbows while keeping them pointed slightly back. Lower down until your chest grazes the floor. Exhale on the way down, then inhale. Maintain a straight spine.

Upper **body**
A push-up targets the **pectoralis major** muscle, **serratus anterior** muscle, and the **deltoids**. In addition, a correct push-up also engages the **triceps brachii** as a secondary muscle, as it's solely responsible for straightening and bending the elbow joint. The **core muscles** work to brace the spine by holding it in a neutral position.

» VARIATIONS

Different push-up variations focus on isolating and targeting various muscle groups. We've selected a few to isolate the triceps, the chest, and the shoulders.

TRICEPS **PUSH-UP**

The triceps push-up is a compound exercise that works muscle groups across your whole body, but specifically isolating the triceps. This variation modifies your hand and arm position and arm path.

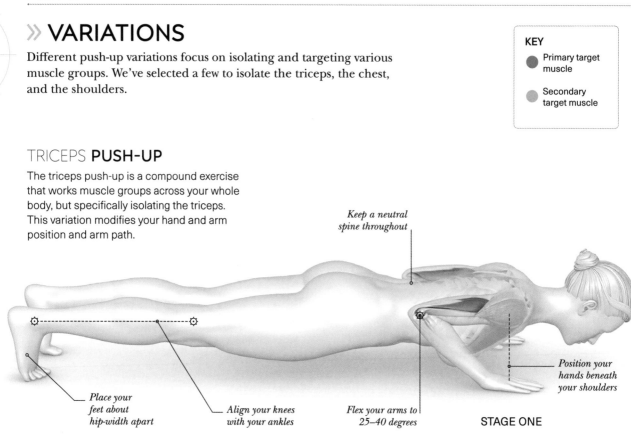

Keep a neutral spine throughout

Position your hands beneath your shoulders

Place your feet about hip-width apart

Align your knees with your ankles

Flex your arms to 25–40 degrees

STAGE ONE

PREPARATORY STAGE
Begin in the high plank position (see pp.36–37), with your hands beneath your shoulders, your neck and spine neutral, and your feet hip-width apart.

STAGE ONE
Engage the core, then inhale as you lower your body toward the floor, flexing your elbows and squeezing your arms against your ribcage.

STAGE TWO
Exhale as you push yourself back up, extending your elbows almost completely to come back into the high plank position. Repeat stages 1 and 2.

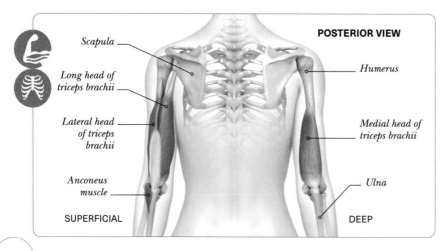

Scapula

Long head of triceps brachii

Lateral head of triceps brachii

Anconeus muscle

SUPERFICIAL

POSTERIOR VIEW

Humerus

Medial head of triceps brachii

Ulna

DEEP

A closer look at **the triceps**
The triceps, also called the triceps brachii, is a large muscle on the back of the arm. It consists of three parts: the lateral and medial heads, which attach at the humerus and in the elbow, and the long head, which attaches at the scapula or shoulder blade. Some triceps movements train all three heads at the same time; others train just one or two. If you know about the anatomy of where each head attaches to the bone, you will be able to make sense of why one exercise can train more triceps muscles than another.

SIDE-TO-SIDE PUSH-UP

Because you alternate sides in this variation, one side at any given point is supporting your entire body. It's important to keep the body tight and controlled. This movement primarily targets the pectoralis major in the chest, while the abdominal muscles act as stabilizers.

Legs stay strong throughout movement

Arms wider than shoulder-width apart

Keep the back flat

Flex the toes

Bend the right elbow

Hands turned out to side

Chest touches floor briefly

PREPARATORY STAGE

STAGE ONE

PREPARATORY STAGE
Start in the high plank position (see pp.36–37) but with arms wider than shoulder-width apart, fingertips turned away from the body. Keep the body straight.

STAGE ONE
Lower down on the right side by bending the right elbow and straightening the left arm out to the left, bringing your chest to the floor briefly. Return to starting point.

STAGE TWO
Lower the body on the left side by bending the left elbow and straightening the right arm out to the right side.

DIAMOND PUSH-UP

This push-up is named from the diamond shape you form with your hands as you complete the exercise. It focuses the weight on the triceps.

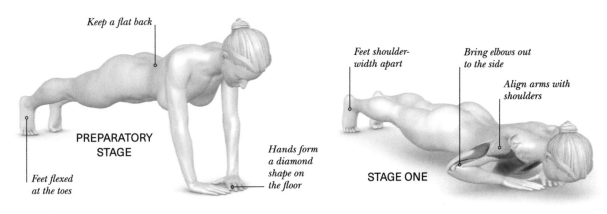

Keep a flat back

Feet shoulder-width apart

Bring elbows out to the side

Align arms with shoulders

PREPARATORY STAGE

Feet flexed at the toes

Hands form a diamond shape on the floor

STAGE ONE

PREPARATORY STAGE
Start in the high plank position (see pp.36–37) with your pelvis tucked in and head and neck neutral. Form a diamond with your hands, under the chest.

STAGE ONE
Engaging the core, slowly bend the elbows so they come out to either side of your body in line with the shoulders. Lower your body to hit the "diamond."

STAGE TWO
Hold for 2 seconds and exhale as you straighten the arms, returning to the starting position. Keep your hands in the diamond shape throughout. Repeat.

OVERHEAD TRICEPS **EXTENSION**

This exercise is very versatile; you can perform it with a dumbbell, a kettlebell, resistance bands, or even a bottle of water. The triceps extension isolates, and in turn strengthens, the muscle on the back of the upper arm. All three "heads" of the triceps—the long head, the lateral head, and the medial head—work to extend the forearm.

THE **BIG PICTURE**

Keep the head aligned over the midline of the chest and the chest aligned over the hips. Maintain a forward focus and don't lower your chin to the chest. It's important that you are utilizing a full range of motion: you should be lowering the weight down to a 90-degree angle and then lifting all the way up. Start with 4 sets of 8 reps; discover variations on the overhead triceps extension on pp.66–67.

KEY

- ● --- *Joints*
- ○— *Muscles*
- ● Shortening with tension
- ● Lengthening with tension
- ● Lengthening without tension
- ● Held muscles without motion

❗ Caution

Make sure you keep the head still and in alignment with the neck and back. You want to isolate the movement to the elbow joint. Another thing that can lead to injury is the placement of the elbows, which should be as close to the ears as possible. Hold the arms directly overhead, biceps by the ears.

Upper **body**

The three muscles of the **triceps** work to stabilize the shoulders and elbows; the shoulders are secondary. At the top of each rep, the **deltoids** contract and shorten. When doing this exercise standing, the **abdominal muscles** are also engaged, being held in an isometric hold.

Triceps
Brachialis
Biceps
Deltoids
Trapezius (upper)
Teres major
Trapezius (lower)
Iliocostalis
Serratus anterior

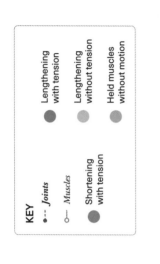

THREE-QUARTER ANTERIOR VIEW

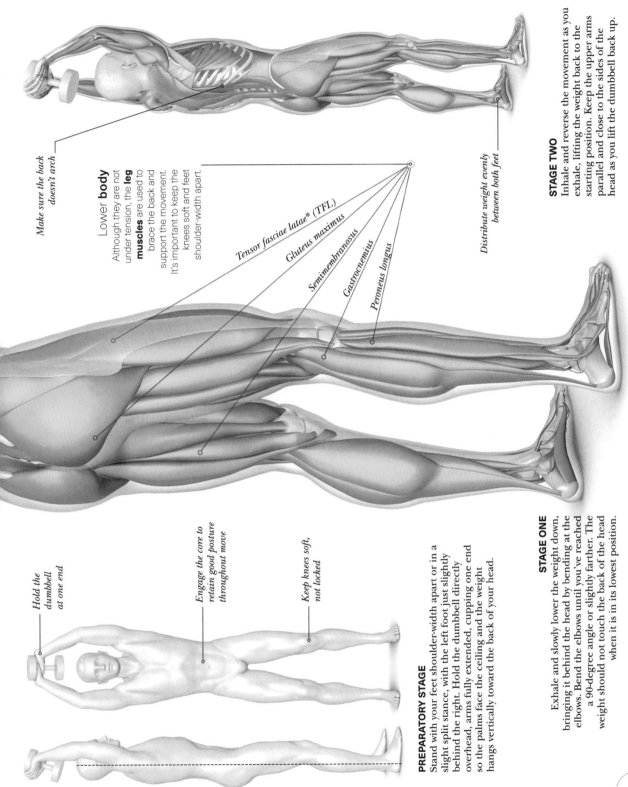

Make sure the back doesn't arch

Lower **body**

Although they are not under tension, the **leg muscles** are used to brace the back and support the movement. It's important to keep the knees soft and feet shoulder-width apart.

Tensor fasciae latae* (TFL)

Gluteus maximus

Semimembranosus

Gastrocnemius

Peroneus longus

Distribute weight evenly between both feet

STAGE TWO

Inhale and reverse the movement as you exhale, lifting the weight back to the starting position. Keep the upper arms parallel and close to the sides of the head as you lift the dumbbell back up.

Hold the dumbbell at one end

Engage the core to retain good posture throughout move

Keep knees soft, not locked

PREPARATORY STAGE

Stand with your feet shoulder-width apart or in a slight split stance, with the left foot just slightly behind the right. Hold the dumbbell directly overhead, arms fully extended, cupping one end so the palms face the ceiling and the weight hangs vertically toward the back of your head.

STAGE ONE

Exhale and slowly lower the weight down, bringing it behind the head by bending at the elbows. Bend the elbows until you've reached a 90-degree angle or slightly farther. The weight should not touch the back of the head when it is in its lowest position.

» VARIATIONS

All of these overhead triceps extension variations work as isolation exercises for the triceps; each one targets all three heads of the muscle. The movements are extremely versatile and can be performed with weights or resistance bands.

*It's important to have a **straight spine** when performing the triceps dip, as a curved spine can add **pressure** to the back.*

TRICEPS **KICKBACK**

The triceps kickback focuses primarily on the lateral head of the triceps. It also targets the abdominals, shoulders, and glutes.

KEY

● Primary target muscle

● Secondary target muscle

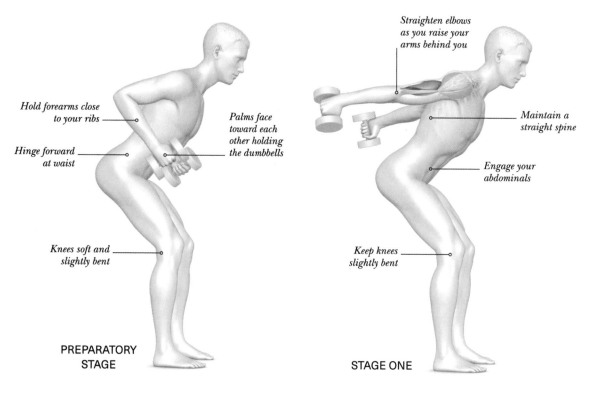

Hold forearms close to your ribs

Hinge forward at waist

Palms face toward each other holding the dumbbells

Knees soft and slightly bent

PREPARATORY STAGE

Straighten elbows as you raise your arms behind you

Maintain a straight spine

Engage your abdominals

Keep knees slightly bent

STAGE ONE

PREPARATORY STAGE
Stand with your feet shoulder-width apart, holding a dumbbell in each hand. Hinge forward at the waist to 45 degrees, pull the elbows up, and hold them at a 90-degree angle.

STAGE ONE
Keeping your head in line with your spine and tucking your chin in slightly, exhale and engage your triceps by straightening your elbows behind you.

STAGE TWO
Hold your upper arms still, only moving your forearms to isolate the triceps. Pause, then inhale as you return the weights to the starting position.

TRICEPS **DIP**

The triceps dip works all three heads of the triceps muscles. When you are accustomed to the movement, you can progress to sitting on the edge of a chair, step, or bench, your hands gripping the edge next to your hips, then slide forward to begin the dips.

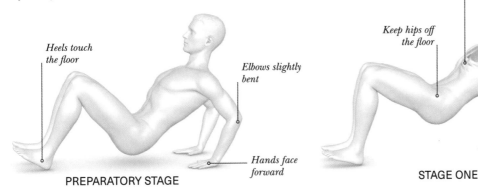

Heels touch the floor

Elbows slightly bent

PREPARATORY STAGE

Hands face forward

Keep spine straight

Keep hips off the floor

Don't lock out the elbows at top of movement

STAGE ONE

PREPARATORY STAGE
Sit on the floor, knees bent and hip-width apart, toes pointed forward, heels on the floor, and palms planted behind you facing the heels.

STAGE ONE
Press into your palms to lift your hips off the floor, using your arms to power the movement. Your elbows will be slightly bent.

STAGE TWO
Slowly lower yourself back down, but don't rest your butt on the floor; go straight back up again. Repeat the movement, using control.

TRICEPS **DIP WITH TOE TOUCH**
(A.K.A. **CRAB** TOUCH)

The is a full-body exercise targeting the glutes, hamstrings, quads, and core. It trains balance, core strength, and many muscle groups, making it work well in a time-efficient body weight circuit workout.

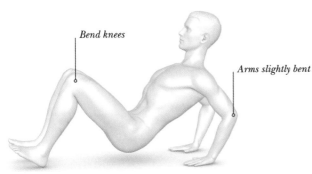

Bend knees

Arms slightly bent

PREPARATORY STAGE

Reach out to touch the opposite toes

Maintain spine alignment when touching toes

As you touch toes, your other arm supports you

Lift hips

STAGE ONE

PREPARATORY STAGE
Sit on the floor in the same position as the triceps dip preparatory pose, knees bent and hands on the floor behind you.

STAGE ONE
Lift your hips, kick your left leg upward, and tap the toes with your right hand, keeping your left hand on the floor.

STAGE TWO
Return the left leg to the floor, go down into a triceps dip, then raise your hips again and repeat the movement on the other side.

DUMBBELL BICEPS CURL

The biceps curl, which you can do seated or standing, isolates and works the biceps muscles at the front of the upper arm, as well as the muscles of the lower arm—the brachialis and brachioradialis.

THE **BIG PICTURE**

By doing this exercise, you build strength in the upper arm and learn to use your arm muscles correctly, bracing the body with your core muscles. Choose the appropriate weights for your fitness levels—if the weights are too heavy, this can cause injuries. Learning how to maintain correct form in the back and the core is engaged will also help prevent strain. Beginners should start with low weights, performing 3–4 sets of 8–10 reps.

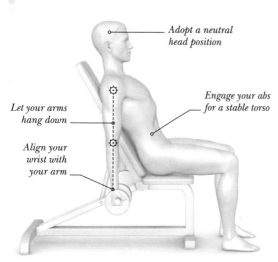

Adopt a neutral head position

Engage your abs for a stable torso

Let your arms hang down

Align your wrist with your arm

PREPARATORY STAGE

Start in a seated position, with your feet and knees hip-width apart. Press your back against the back of your chair. Hold dumbbells with an underhand grip, letting your arms relax down at the sides of your body with palms facing forward and shoulders back.

Deltoids
Triceps
Brachialis
Biceps
Brachioradialis
Extensor digitorum

Arms

The muscles recruited for a biceps curl include the **anterior deltoid**, the **biceps brachii**, the **brachialis muscle**, the **brachioradialis**, and the **flexors** and **extensors** of the forearm. Most of these muscles stabilize the shoulder, wrist, and elbow during the curl, and the **muscles of the forearm** control grip strength.

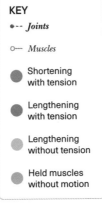

Sternocleidomastoid
Trapezius
Pectoralis major
Serratus anterior
Latissimus dorsi
Transversus abdominis

Upper **body**

The muscles in the **abdominals** brace the back, creating stability while lifting. The **upper back** works to hold the abdominals but also hold the spine, neck, and head in neutral. Always keep the core engaged to keep your back in contact with the chair or bench.

ANTERIOR-LATERAL VIEW

Stabilize your shoulder blades by activating your upper back muscles

Press your back against the seat pad

Keep your hips and lower back still at all times

Distribute your weight evenly through your feet

STAGE TWO

At the top of the curl, hold the tension in the arm for 2 seconds, then slowly and with control lower your weight back to the starting position. Don't swing the weight when lowering. Reset, and repeat stages 1 and 2.

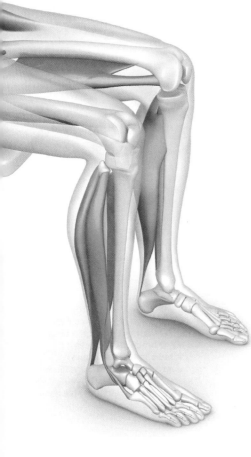

KEY

●-- *Joints*

○— *Muscles*

● Shortening with tension

● Lengthening with tension

● Lengthening without tension

● Held muscles without motion

STAGE ONE

Keeping your upper arms stable and shoulders relaxed, exhale, bend at the elbow, and lift the weights so that the dumbbells approach your shoulders. Your elbows should stay tucked in close to your ribs.

❗ Common mistakes

Make sure, when lifting, you choose appropriate weights for your fitness level. If the weights are too heavy, the body will contort to push the weights up. Make sure you aren't lifting the weights up too fast, as that can add stress to the elbows.

» VARIATIONS

These biceps curl variations look to specifically target different muscles in the biceps. Hammer curls target the long head of the biceps, as well as the brachialis and the brachioradialis. The wide curl targets the inside of your biceps, also known as the short head.

WIDE BICEPS CURL

The wide biceps curl is an isolation exercise that strengthens the biceps. A wide grip places more emphasis on the short head of the biceps brachii. In addition to the biceps, secondary muscles worked are the deltoids and the abdominals.

> *Choose the **appropriate weights** for your fitness level. If the weights are too heavy, this can cause injuries.*

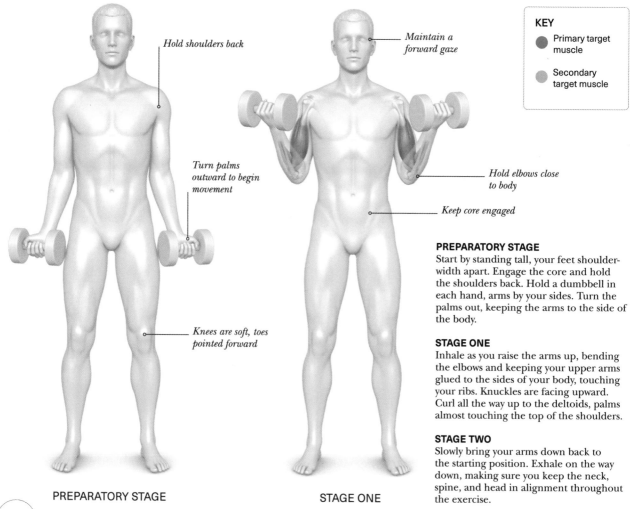

Hold shoulders back

Turn palms outward to begin movement

Knees are soft, toes pointed forward

Maintain a forward gaze

Hold elbows close to body

Keep core engaged

KEY
- Primary target muscle
- Secondary target muscle

PREPARATORY STAGE

STAGE ONE

PREPARATORY STAGE
Start by standing tall, your feet shoulder-width apart. Engage the core and hold the shoulders back. Hold a dumbbell in each hand, arms by your sides. Turn the palms out, keeping the arms to the side of the body.

STAGE ONE
Inhale as you raise the arms up, bending the elbows and keeping your upper arms glued to the sides of your body, touching your ribs. Knuckles are facing upward. Curl all the way up to the deltoids, palms almost touching the top of the shoulders.

STAGE TWO
Slowly bring your arms down back to the starting position. Exhale on the way down, making sure you keep the neck, spine, and head in alignment throughout the exercise.

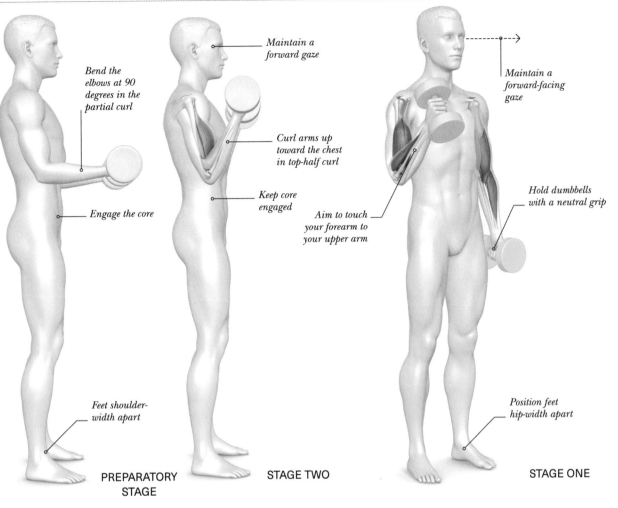

Bend the
elbows at 90
degrees in the
partial curl

Engage the core

Feet shoulder-
width apart

**PREPARATORY
STAGE**

Maintain a
forward gaze

Curl arms up
toward the chest
in top-half curl

Keep core
engaged

Aim to touch
your forearm to
your upper arm

STAGE TWO

Maintain a
forward-facing
gaze

Hold dumbbells
with a neutral grip

Position feet
hip-width apart

STAGE ONE

PARTIAL BICEPS CURL

This is a biceps isolation exercise. You can either do partial curls at the bottom half, or at the top half of a full biceps curl. Both moves strengthen the biceps muscles at the front of the upper arm and the lower back muscles.

PREPARATORY STAGE
Start by holding a dumbbell in each hand in front of you, knuckles facing out, feet shoulder-width apart. Keep your knees soft, shoulders back, and head and spine neutral.

STAGE ONE
Bend at the elbow and lift the weights to a 90-degree angle. Tuck in your elbows close to your ribs. Pause here.

STAGE TWO
Slowly lower your arms back to the starting position. To vary the exercise, begin with the arms at the 90-degree angle, then curl up to the shoulders and return the weights back to the midway point.

HAMMER CURL

As well as the biceps, this variation also challenges other elbow flexors—the brachioradialis and brachialis. You can perform it using one arm at a time (as above) or both arms together. Hold for 1–2 seconds at the top position.

PREPARATORY STAGE
Hold a dumbbell in each hand and stand tall with your arms hanging at your sides. Keep your wrists relaxed when holding the weights.

STAGE ONE
Inhale and engage your core, then exhale as you curl up (either with one arm or both), flexing your elbow to the top position.

STAGE TWO
Inhale as you lower your arm(s). Repeat stages 1 and 2, being sure to do the same reps on either arm if lifting one arm at a time.

DUMBBELL
FRONT RAISE

The front raise is an isolation exercise for the deltoids (shoulder muscles), which are the primary muscles used here. It also works the pectorals (upper chest muscles).

THE **BIG PICTURE**

The dumbbell front raise is a great exercise for beginners. Start by choosing dumbbells of a suitable weight. If you're new to weight lifting, you should start with a set of light weights and plan to do 10–12 reps for 3 sets. As you lift the weights, think 3 counts up and 3 counts down, controlling the weights all the way up and down.

Common mistakes

When performing this exercise, it's important not to rock or sway—if you need do this to lift the weights, they are probably too heavy. To help prevent this action, it's important to keep the abdominals tight, as they aid in bracing the back, keeping it straight and in the neutral position.

Wrist

F.d. superficialis

Brachioradialis

Pronator teres

Elbow

Biceps

Deltoids

Triceps

Pectoralis major

Latissimus dorsi

Serratus anterior

Spinal extensors

Psoas major

Transversus abdominis

Upper body and **arms**

This exercise targets the **anterior** and **medial deltoids**, which sit on the front and sides of the shoulder. Secondary muscles that aid in the lift are the **trapezius, erector spinae, biceps, pectorals, rotator cuff, serratus anterior,** and **abdominals.**

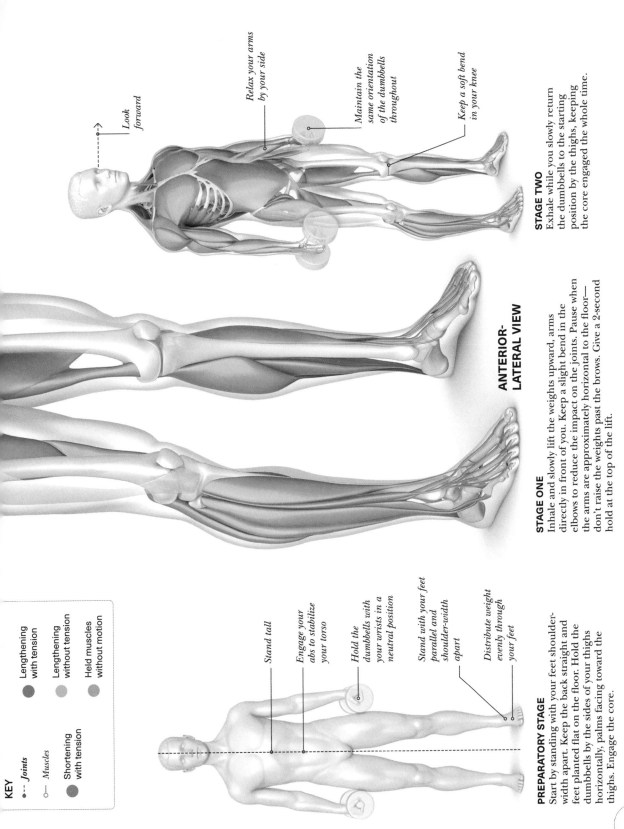

KEY

- **- - - Joints**
- **—o— Muscles**
- ● Shortening with tension
- ● Lengthening with tension
- ● Lengthening without tension
- ● Held muscles without motion

Look forward

Relax your arms by your side

Maintain the same orientation of the dumbbells throughout

Keep a soft bend in your knee

ANTERIOR-LATERAL VIEW

STAGE ONE

Inhale and slowly lift the weights upward, arms directly in front of you. Keep a slight bend in the elbows to reduce the impact on the joints. Pause when the arms are approximately horizontal to the floor—don't raise the weights past the brows. Give a 2-second hold at the top of the lift.

STAGE TWO

Exhale while you slowly return the dumbbells to the starting position by the thighs, keeping the core engaged the whole time.

Stand tall

Engage your abs to stabilize your torso

Hold the dumbbells with your wrists in a neutral position

Stand with your feet parallel and shoulder-width apart

Distribute weight evenly through your feet

PREPARATORY STAGE

Start by standing with your feet shoulder-width apart. Keep the back straight and feet planted flat on the floor. Hold the dumbbells by the sides of your thighs horizontally, palms facing toward the thighs. Engage the core.

» VARIATIONS

Exercises that employ the rowing movement primarily target the muscles on your back, but also improve core stability, while engaging muscles on your shoulders and arms. The main muscles worked are: latissimus dorsi, rhomboids, trapezius, rear deltoids, erector spinae, and the biceps.

DUMBBELL **BENT-OVER ROW**

When using the dumbbells, you can row unilaterally, performing the move with one leg supported on a bench, or bilaterally, with both feet planted shoulder-width apart and hips flexed at a 90-degree angle. Hold for 2 seconds at the top of the lift to take it up a notch.

PREPARATORY STAGE
Place your opposite knee on the bench and position the other leg under your hip. Make sure the hips are squared up to the floor and one side isn't lifted. Your back should be flat, head in alignment with the spine and neck. Breathe deeply to activate the core, bracing the back.

STAGE ONE
Breathe out as you retract the shoulder blades and raise the arm. Flex the elbow between 30 and 75 degrees to change muscle bias.

STAGE TWO
Breathe in as you lower the dumbbell in a controlled movement. Make sure you keep the core engaged. Repeat stages 1 and 2.

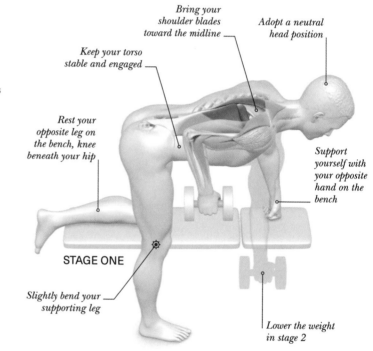

Bring your shoulder blades toward the midline

Adopt a neutral head position

Keep your torso stable and engaged

Rest your opposite leg on the bench, knee beneath your hip

Support yourself with your opposite hand on the bench

STAGE ONE

Slightly bend your supporting leg

Lower the weight in stage 2

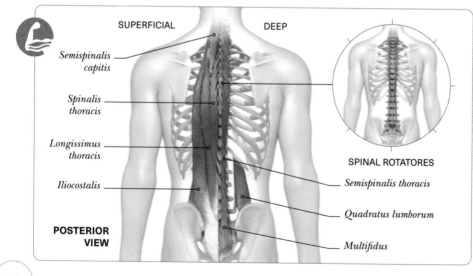

SUPERFICIAL   DEEP

Semispinalis capitis

Spinalis thoracis

Longissimus thoracis

Iliocostalis

POSTERIOR VIEW

SPINAL ROTATORES

Semispinalis thoracis

Quadratus lumborum

Multifidus

Spinal **extensors**
The extensor muscles are attached to the back of the spine and enable standing and lifting. These include the outer large paired muscles extending the vertebral column, collectively known as erector spinae, which help hold up the spine. Deeper muscles, including the rotatores, support the work of the erectors, also stabilizing the pelvis. Strengthening these muscles aids in supporting the body, improves posture, and can help alleviate lower back pain.

KEY

● Primary target muscle

● Secondary target muscle

BANDED UPRIGHT ROW

This variation uses a band for resistance. Bands come in a variety of resistance sizes; choose the best for your current fitness level. Hold for 2 seconds at the top of the lift.

Working the **upper traps**

The upper traps (trapezius muscles) support your arms and raise your shoulder blades, while mid and lower traps are essential for scapular retraction, depression, and rotation. Shrugging your shoulders, raising your arms, and other movements use your traps; it's important not to neglect them.

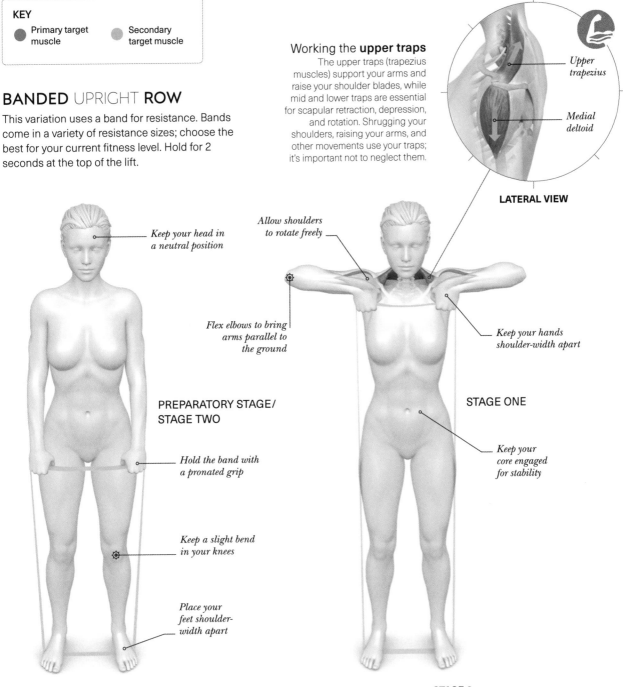

Upper trapezius

Medial deltoid

LATERAL VIEW

Keep your head in a neutral position

Allow shoulders to rotate freely

Flex elbows to bring arms parallel to the ground

Keep your hands shoulder-width apart

PREPARATORY STAGE/ STAGE TWO

STAGE ONE

Hold the band with a pronated grip

Keep your core engaged for stability

Keep a slight bend in your knees

Place your feet shoulder-width apart

PREPARATORY STAGE
Place the resistance band beneath the feet and stand straight up. Hold the band just below the hips, feet and hands shoulder-width apart, with a soft bend in the knees.

STAGE ONE
Inhale to activate the abdominals. As you exhale, raise your shoulders up toward the ceiling, bringing your hands upward and flexing the elbows.

STAGE 2
Inhale as you slowly lower your shoulders and extend your arms back to the starting position. Repeat stages 1 and 2.

DUMBBELL LATERAL RAISE

> **❗ Common mistakes**
>
> If weights are too heavy, momentum will be needed to get the weights up, which will cause you to swing the weights with a jerking movement. This doesn't isolate the shoulders and can create injuries. Make sure you keep a neutral spine throughout.

The lateral (side) raise primarily works the lateral deltoid, which is the middle portion of the deltoid muscle. The anterior (front) deltoid, posterior (back) deltoid, upper trap, supraspinatus (a rotator cuff muscle), and serratus anterior (muscles along your ribs under your armpit) also contribute to the movement. If performed regularly, this can help you build broad shoulders.

THE BIG PICTURE

Make sure you engage the core and raise and lower your arms slowly and with control—don't just let the weights fall downward. If you're new to weight lifting, you should start with a set of light weights and plan to do 10–12 reps for 3 sets.

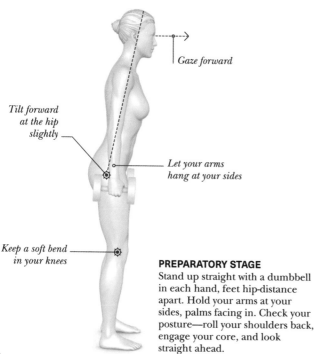

Gaze forward

Tilt forward at the hip slightly

Let your arms hang at your sides

Keep a soft bend in your knees

PREPARATORY STAGE
Stand up straight with a dumbbell in each hand, feet hip-distance apart. Hold your arms at your sides, palms facing in. Check your posture—roll your shoulders back, engage your core, and look straight ahead.

Wrist
Extensor digitorum
Flexor digitorum superficialis
Brachioradialis
Flexor digitorum profundus
Elbow
Biceps
Triceps
Deltoids
Supraspinatus
Infraspinatus
Teres major
Serratus anterior
Trapezius
Transversus ab.
Spinal extensors
Spine

Upper **body** and **arms**
The **anterior deltoid**, the **supraspinatus**, and the **trapezius** muscles assist the **lateral deltoids** during this exercise. The anterior deltoids are located in the front of your shoulders. The supraspinatus in the **rear deltoid** initiates the movement. Your trapezius is responsible for the shoulder elevation.

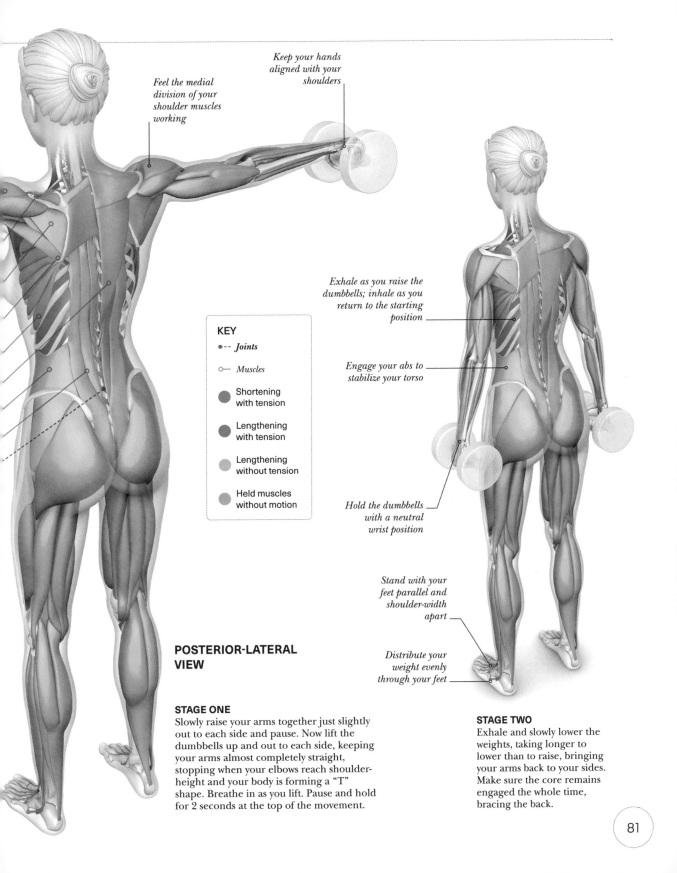

Feel the medial division of your shoulder muscles working

Keep your hands aligned with your shoulders

Exhale as you raise the dumbbells; inhale as you return to the starting position

Engage your abs to stabilize your torso

Hold the dumbbells with a neutral wrist position

Stand with your feet parallel and shoulder-width apart

Distribute your weight evenly through your feet

KEY

●-- *Joints*

○— *Muscles*

● Shortening with tension

● Lengthening with tension

● Lengthening without tension

● Held muscles without motion

POSTERIOR-LATERAL VIEW

STAGE ONE

Slowly raise your arms together just slightly out to each side and pause. Now lift the dumbbells up and out to each side, keeping your arms almost completely straight, stopping when your elbows reach shoulder-height and your body is forming a "T" shape. Breathe in as you lift. Pause and hold for 2 seconds at the top of the movement.

STAGE TWO

Exhale and slowly lower the weights, taking longer to lower than to raise, bringing your arms back to your sides. Make sure the core remains engaged the whole time, bracing the back.

MILITARY SHOULDER PRESS

This exercise strengthens the pectorals (chest), deltoids (shoulders), triceps (arms), and trapezius (upper back). Standing upright requires balance, so you also use your core muscles and lower back muscles. You can also try this exercise seated.

THE **BIG PICTURE**

When raising dumbbells overhead, keep your elbows either directly underneath your wrists or slightly more inward. Avoid locking the elbows out. Tightening the core and glutes helps stabilize the spine when lifting weights above your head. If you're a beginner, start with 1 set of 8–10 reps.

Caution

If the load is too heavy, it can cause your lower back to arch and create lower back pain. Also take care when picking up your weights—flex at the knees and waist to reach for them.

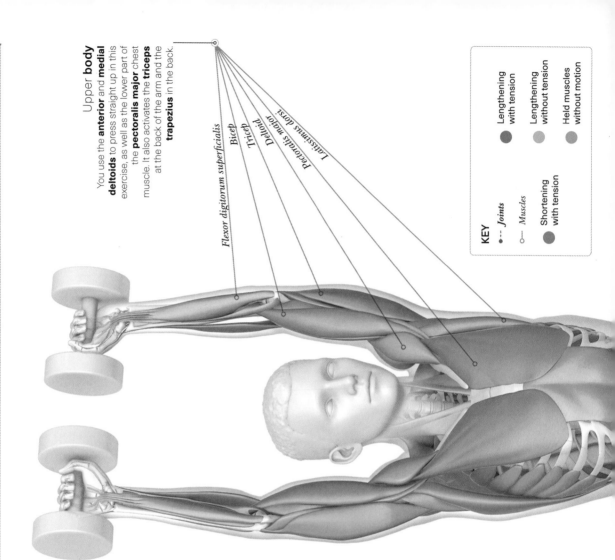

Upper **body**

You use the **anterior** and **medial deltoids** to press straight up in this exercise, as well as the lower part of the **pectoralis major** chest muscle. It also activates the **triceps** at the back of the arm and the **trapezius** in the back.

Flexor digitorum superficialis

Bicep

Tricep

Deltoid

Pectoralis major

Latissimus dorsi

KEY

• - - *Joints*

o— *Muscles*

● Lengthening with tension

● Lengthening without tension

● Held muscles without motion

● Shortening with tension

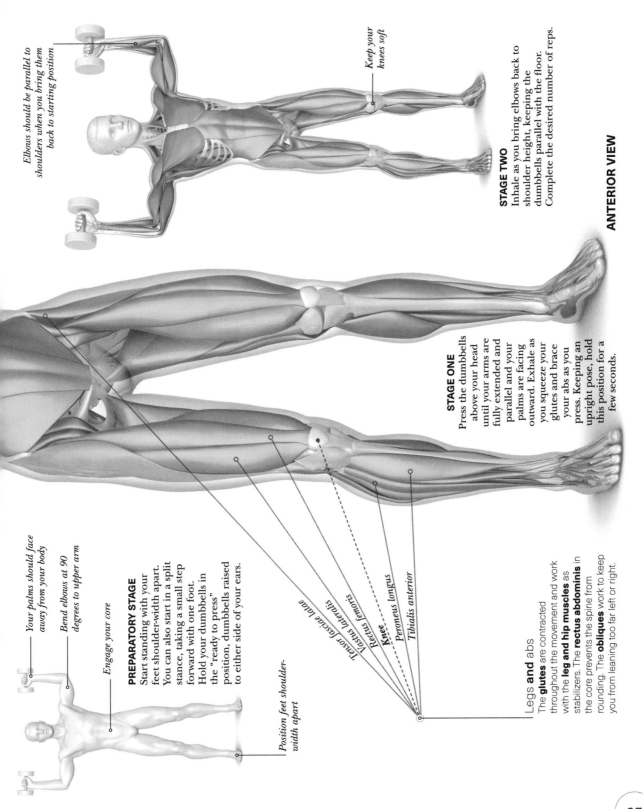

Elbows should be parallel to shoulders when you bring them back to starting position

Keep your knees soft

STAGE TWO
Inhale as you bring elbows back to shoulder height, keeping the dumbbells parallel with the floor. Complete the desired number of reps.

ANTERIOR VIEW

STAGE ONE
Press the dumbbells above your head until your arms are fully extended and your palms are facing outward. Exhale as you squeeze your glutes and brace your abs as you press. Keeping an upright pose, hold this position for a few seconds.

Your palms should face away from your body

Bend elbows at 90 degrees to upper arm

Engage your core

PREPARATORY STAGE
Start standing with your feet shoulder-width apart. You can also start in a split stance, taking a small step forward with one foot. Hold your dumbbells in the "ready to press" position, dumbbells raised to either side of your ears.

Position feet shoulder-width apart

Tensor fasciae latae

Vastus lateralis

Rectus femoris

Knee

Peroneus longus

Tibialis anterior

Legs **and** abs
The **glutes** are contracted throughout the movement and work with the **leg and hip muscles** as stabilizers. The **rectus abdominis** in the core prevents the spine from rounding. The **obliques** work to keep you from leaning too far left or right.

83

» VARIATIONS

The primary muscles used in shoulder presses are the deltoids, which consist of the anterior (front), medial (middle), and posterior (back) deltoid. These variations focus on different parts of the deltoid. The neutral grip overhead press targets mostly the anterior head of the deltoids, as well as the medial head. The Arnold press, named after Arnold Schwarzenegger, focuses on all three of the shoulder heads at the same time.

KEY
- ● Primary target muscle
- ● Secondary target muscle

NEUTRAL-GRIP DUMBBELL SHOULDER PRESS

This variation isolates the anterior head of the deltoids in addition to the medial head. By having the palms facing each other (in a neutral grip), instead of facing outward, this angle targets different muscles within the deltoids.

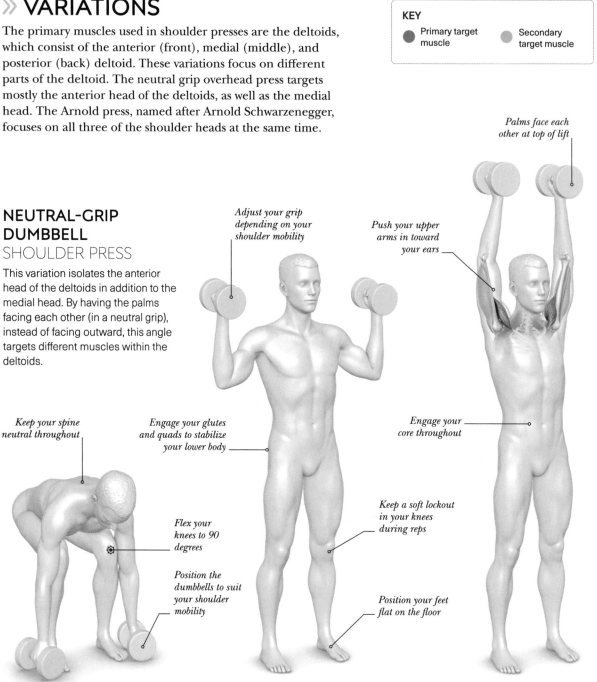

Palms face each other at top of lift

Adjust your grip depending on your shoulder mobility

Push your upper arms in toward your ears

Keep your spine neutral throughout

Engage your glutes and quads to stabilize your lower body

Engage your core throughout

Flex your knees to 90 degrees

Keep a soft lockout in your knees during reps

Position the dumbbells to suit your shoulder mobility

Position your feet flat on the floor

PICKING UP THE WEIGHTS SAFELY
Stand with your feet hip- to shoulder-width apart. Flex at the knees and hips to reach the dumbbells, which should be at either side of your feet.

PREPARATORY STAGE/STAGE TWO
Straighten your knees and raise the dumbbells to above shoulder height, in the "ready to press" position. Keep your core engaged as you prepare to press up.

STAGE ONE
Inhale to brace your abdominals, then exhale as you drive the dumbbells upward, palms facing each other. Inhale to return to stage 2. Repeat both stages.

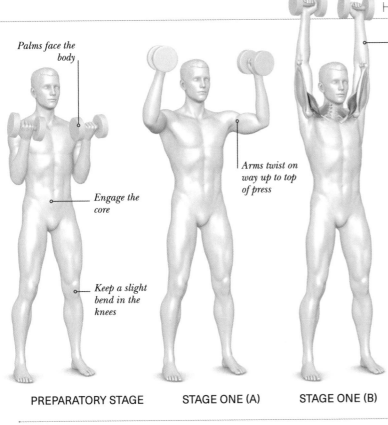

Palms face the body

Engage the core

Keep a slight bend in the knees

Arms twist on way up to top of press

Press upward to extend the arms

PREPARATORY STAGE

STAGE ONE (A)

STAGE ONE (B)

ARNOLD PRESS

This press hits all three sections of the deltoid, the round-looking muscle that caps the top of your upper arm—the front deltoid, medial deltoid, and back deltoid. It's a good exercise to strengthen shoulders.

PREPARATORY STAGE
Hold a pair of dumbbells at shoulder height, palms facing your body, and stand straight with your feet around shoulder-width apart and your knees soft.

STAGE ONE
Raise the dumbbells to just above shoulder height on each side, twisting so that your palms are parallel to your body. Slowly raise the dumbbells above your head while turning your wrists so that your palms face forward, until your arms are fully extended.

STAGE TWO
Do not pause at the top of the movement. Lower the dumbbells back to the starting position, twisting until your palms are facing your body once again. Repeat as desired.

BODY WEIGHT INVERTED SHOULDER PRESS

The floor/bench inverted shoulder press is a variation to the basic push-up that increases strength in the chest, shoulders, and triceps. The inverted angle places more emphasis on the shoulders and triceps and less emphasis on the chest.

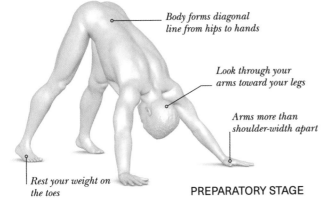

Body forms diagonal line from hips to hands

Look through your arms toward your legs

Arms more than shoulder-width apart

Rest your weight on the toes

PREPARATORY STAGE

Hips remain in air for duration of press

Bend elbows at a 90-degree angle

Toes remain on floor during press

STAGE ONE

PREPARATORY STAGE
Place your hands far apart on the floor and raise your hips up so your body forms an inverted "V."

STAGE ONE
Bend your elbows to 90 degrees and lower your upper body until your head nearly touches the floor.

STAGE TWO
Exhale and press back up to the starting position. Keep the crown of your head facing the floor.

DUMBBELL **REAR** DELTOID FLY

> **!** **Common mistakes**
> Using weights that are too heavy can lead to hunching or rounding of the back, placing stress on the spine. Keep your chin tucked to maintain a neutral position in the spine, and engage the core.

The reverse fly exercise targets the rear shoulders (deltoids) and major muscles of the upper back, including the trapezius. The trapezius helps with a scapular retraction (pulling your shoulder blades in toward each other). Strengthening these muscles will help improve poor posture, promote an upright stance, and improve balance.

THE **BIG PICTURE**

You may want to practice without weights first, then grab light weights when you're ready to try out the full movement. Move the weights up and down with control—don't throw them out and let them fall back. If you're new to weight lifting, you should start with a set of light weights and plan to do 10–12 repetitions for 3 sets to start with.

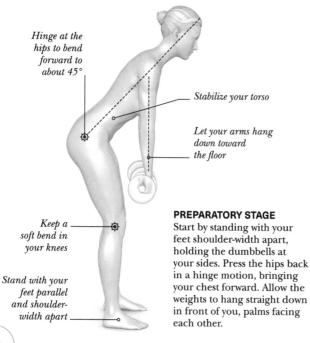

Hinge at the hips to bend forward to about 45°

Stabilize your torso

Let your arms hang down toward the floor

Keep a soft bend in your knees

Stand with your feet parallel and shoulder-width apart

PREPARATORY STAGE
Start by standing with your feet shoulder-width apart, holding the dumbbells at your sides. Press the hips back in a hinge motion, bringing your chest forward. Allow the weights to hang straight down in front of you, palms facing each other.

Flexor digitorum superficialis
Flexor digitorum profundus
Extensor digiti minimi
Extensor digitorum
Brachioradialis
Triceps (medial head)
Triceps (short and long heads)
Deltoids
Sternocleidomastoid
Teres major
Infraspinatus
Trapezius
Serratus anterior
Spinal extensors
Latissimus dorsi
Transversus abdom
Spine

Upper **body** and **arms**
This exercise places tension on the **posterior deltoid, rhomboid**, and **middle trapezius** muscles. The rhomboid muscles, which are located in the **upper back** and **shoulders**, are the primary muscles used during the reverse fly. Make sure you are using proper form and a weight you can control when performing this exercise.

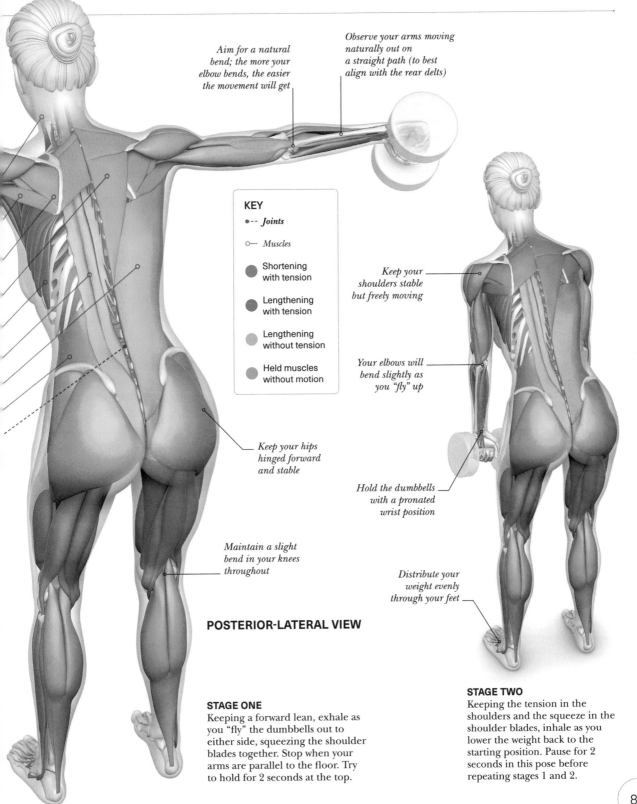

Aim for a natural bend; the more your elbow bends, the easier the movement will get

Observe your arms moving naturally out on a straight path (to best align with the rear delts)

KEY

●-- *Joints*

○— *Muscles*

● Shortening with tension

● Lengthening with tension

○ Lengthening without tension

● Held muscles without motion

Keep your shoulders stable but freely moving

Keep your hips hinged forward and stable

Your elbows will bend slightly as you "fly" up

Hold the dumbbells with a pronated wrist position

Maintain a slight bend in your knees throughout

Distribute your weight evenly through your feet

POSTERIOR-LATERAL VIEW

STAGE ONE
Keeping a forward lean, exhale as you "fly" the dumbbells out to either side, squeezing the shoulder blades together. Stop when your arms are parallel to the floor. Try to hold for 2 seconds at the top.

STAGE TWO
Keeping the tension in the shoulders and the squeeze in the shoulder blades, inhale as you lower the weight back to the starting position. Pause for 2 seconds in this pose before repeating stages 1 and 2.

» VARIATIONS

These reverse fly variations are intended to target different areas of the back. Wide-grip rows recruit the trapezoid, rhomboid, and rear deltoid muscles. In addition, it also works the latissimus dorsi. The pullovers isolate the latissimus dorsi and the pectoralis major. Both variations also isolate the abdominal muscles, which remain engaged throughout.

BENT-OVER WIDE ROW

This row targets muscles in the upper and middle back, as well as those in the upper arm and the shoulder rotators. The upper back muscles help pull the shoulders down and back. Training all these muscles will improve the symmetry of your upper body and will also help you stand up tall and straight and maintain correct posture.

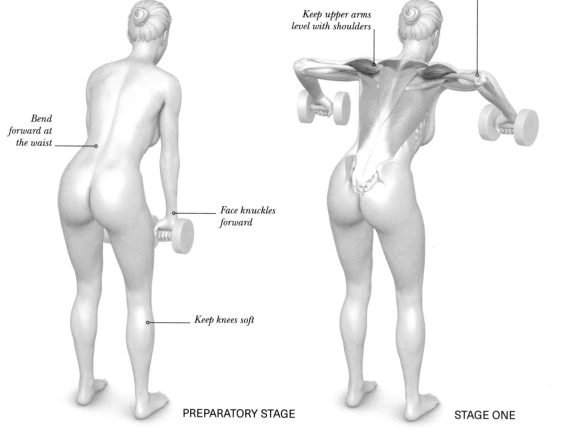

Bend forward at the waist

Face knuckles forward

Keep knees soft

Keep upper arms level with shoulders

Arms at a 90-degree angle, elbows bent

PREPARATORY STAGE

STAGE ONE

PREPARATORY STAGE
Stand up straight, feet shoulder-width apart and core engaged. Hold a dumbbell in each hand in front of your thighs, knuckles facing out. Bend forward.

STAGE ONE
Pull the dumbbells toward your chest, keeping your arms wider than shoulder-width apart, upper arms level with your shoulders. Squeeze the shoulder blades.

STAGE TWO
Lower the weight and return to the starting position: that's 1 rep. Continue completing rows for the rep time.

DUMBBELL PULLOVERS

The basic dumbbell pullover strengthens the muscles in the chest (pectoralis major) and your "wing" muscles in the back (latissimus dorsi). By making variations to the movement, you can also engage the core muscles and the back of the upper arm (triceps). This is a great exercise because it simultaneously trains both the anterior and posterior of the body.

*Core stability helps **protect** your back from injury. If you struggle to keep the core engaged, you may be **lifting too much** weight.*

Palms face each other

Bend the knees

Elbows slightly bent

Plant feet firmly on the floor

Push your back against the floor

Keep knees bent

Extend arms behind your head

Maintain strong wrists

Keep palms facing each other

Engage core throughout movement

Elbows soft, not locked

PREPARATORY STAGE
Lie on the floor, feet hip-width apart and knees bent. Press your lower back against the floor and hold a pair of dumbbells above your chest.

STAGE ONE
Keep a strong back and core while you inhale and extend the weights back and over your head, until they are on the floor behind your head.

STAGE TWO
Once you reach full range of motion, exhale slowly and return your arms to the starting position. Repeat for 30–60 seconds.

DUMBBELL BENCH PRESS

The dumbbell chest press works the pectoral muscles of the chest. It also targets the anterior deltoids of the shoulder, the triceps brachii of the upper arm, the forearms, and the abdominals.

THE BIG PICTURE

This move can easily be done on the floor or on a bench. On the floor, your feet should be flat on the floor hip-distance apart, knees bent. Your back should be pushed against the floor or bench, your core engaged. Push the weights up and bring them down slowly and with control. Keep your body and legs still and strong as you lower and lift. If you're new to weight lifting, you should start with a set of light weights and plan to do 10–12 reps for 3 sets to start with.

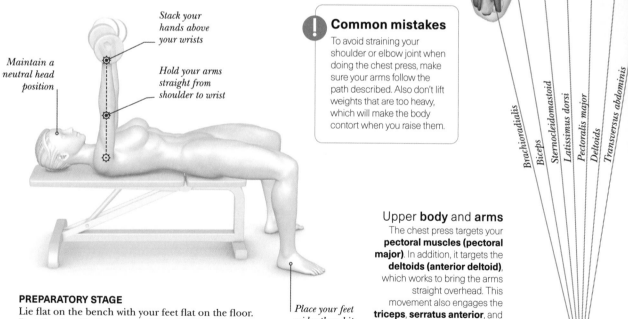

Stack your hands above your wrists

Maintain a neutral head position

Hold your arms straight from shoulder to wrist

⚠ Common mistakes

To avoid straining your shoulder or elbow joint when doing the chest press, make sure your arms follow the path described. Also don't lift weights that are too heavy, which will make the body contort when you raise them.

Brachioradialis
Biceps
Sternocleidomastoid
Latissimus dorsi
Pectoralis major
Deltoids
Transversus abdominis

PREPARATORY STAGE
Lie flat on the bench with your feet flat on the floor. Rest the dumbbells on your thighs, holding them with an overhand grip. Slowly push the weights upward while exhaling; make sure to never lock out the elbows. Your arms should form a straight line.

Place your feet wider than hip-width apart

Upper **body** and **arms**
The chest press targets your **pectoral muscles (pectoral major)**. In addition, it targets the **deltoids (anterior deltoid)**, which works to bring the arms straight overhead. This movement also engages the **triceps**, **serratus anterior**, and **biceps**. The **abdominals** are also used in this movement to brace the spine and maintain stillness in the torso.

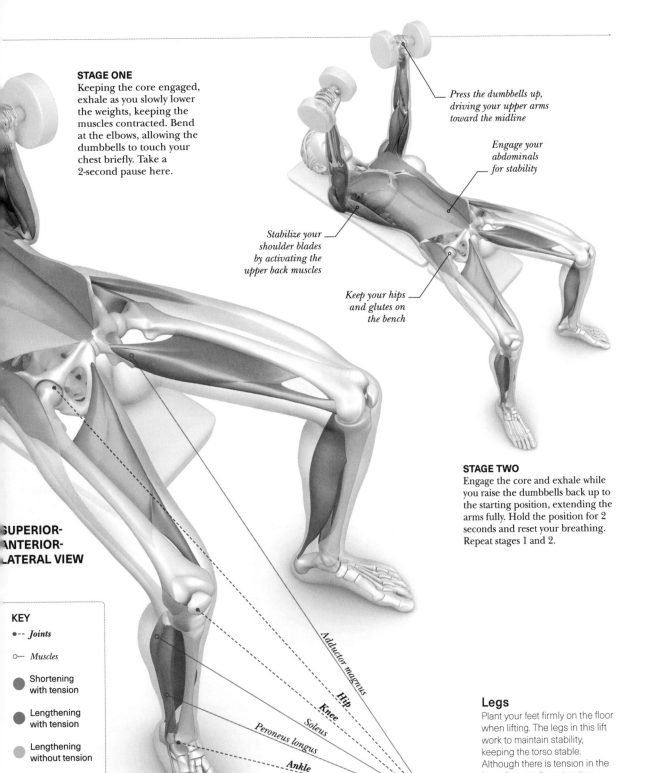

STAGE ONE
Keeping the core engaged, exhale as you slowly lower the weights, keeping the muscles contracted. Bend at the elbows, allowing the dumbbells to touch your chest briefly. Take a 2-second pause here.

Press the dumbbells up, driving your upper arms toward the midline

Engage your abdominals for stability

Stabilize your shoulder blades by activating the upper back muscles

Keep your hips and glutes on the bench

**SUPERIOR-
ANTERIOR-
LATERAL VIEW**

KEY

●--- *Joints*

○--- *Muscles*

● Shortening with tension

● Lengthening with tension

● Lengthening without tension

● Held muscles without motion

Adductor magnus

Hip

Knee

Soleus

Peroneus longus

Ankle

STAGE TWO
Engage the core and exhale while you raise the dumbbells back up to the starting position, extending the arms fully. Hold the position for 2 seconds and reset your breathing. Repeat stages 1 and 2.

Legs
Plant your feet firmly on the floor when lifting. The legs in this lift work to maintain stability, keeping the torso stable. Although there is tension in the legs, they aren't contracting—the **glutes**, **quadriceps**, and **calves** are isometrically held.

91

DUMBBELL
CHEST FLY

This single-joint movement exercise isolates the chest muscles and secondarily works the deltoids, triceps, and the biceps. The movement opens up the chest muscles, which may aid in alleviating chest tightness and can improve posture.

THE **BIG PICTURE**

It's very important when performing a chest fly to practice the correct technique. You should maintain a slow, controlled lift and lower the weights slowly to avoid muscle and joint sprain. If you're new to working with weights, you should start with a set of light weights and plan to do 10–12 reps for 3 sets.

STAGE ONE
Maintain an engaged core. Inhale and, while keeping your elbows slightly bent, exhale and slowly lower the dumbbells directly out to your sides until you feel a deep stretch in your chest. Hold at the bottom for about 2 seconds.

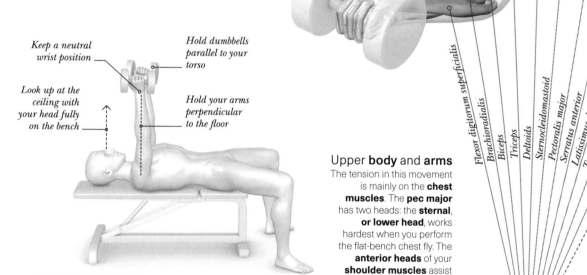

Keep a neutral wrist position

Hold dumbbells parallel to your torso

Look up at the ceiling with your head fully on the bench

Hold your arms perpendicular to the floor

Flexor digitorum superficialis
Brachioradialis
Biceps
Triceps
Deltoids
Sternocleidomastoid
Pectoralis major
Serratus anterior
Latissimus dorsi
Transversus abdominis
Spinal extensors
Spine

PREPARATORY STAGE
Start by lying back on a flat bench. (You can also do this lying on the floor.) Plant your feet firmly on the floor, feet shoulder-width apart. Hold a pair of dumbbells at arm's length over your chest, palms facing each other. Keep your head, neck, and spine in a neutral position.

Upper **body and arms**
The tension in this movement is mainly on the **chest muscles**. The **pec major** has two heads: the **sternal, or lower head**, works hardest when you perform the flat-bench chest fly. The **anterior heads** of your **shoulder muscles** assist your pectorals in this movement. The **biceps** contract isometrically during flys, stabilizing your shoulder joint and **forearm** as you lower the weights.

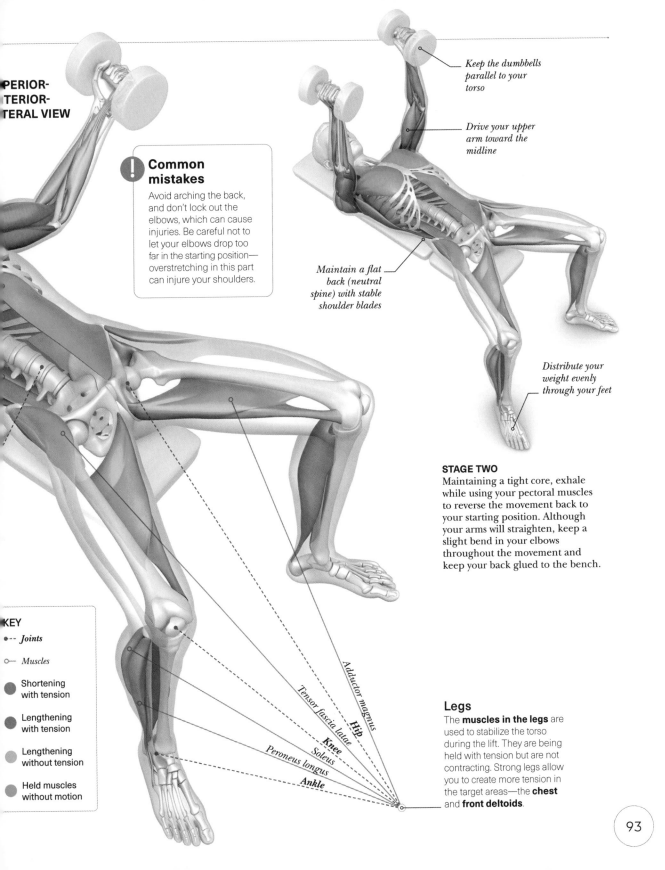

Keep the dumbbells parallel to your torso

Drive your upper arm toward the midline

! Common mistakes

Avoid arching the back, and don't lock out the elbows, which can cause injuries. Be careful not to let your elbows drop too far in the starting position—overstretching in this part can injure your shoulders.

Maintain a flat back (neutral spine) with stable shoulder blades

Distribute your weight evenly through your feet

STAGE TWO

Maintaining a tight core, exhale while using your pectoral muscles to reverse the movement back to your starting position. Although your arms will straighten, keep a slight bend in your elbows throughout the movement and keep your back glued to the bench.

KEY

- •-- *Joints*
- ○— *Muscles*
- ● Shortening with tension
- ● Lengthening with tension
- ○ Lengthening without tension
- ● Held muscles without motion

Adductor magnus
Hip
Tensor fascia latae
Knee
Soleus
Peroneus longus
Ankle

Legs

The **muscles in the legs** are used to stabilize the torso during the lift. They are being held with tension but are not contracting. Strong legs allow you to create more tension in the target areas—the **chest** and **front deltoids**.

93

LOWER BODY
EXERCISES

Exercises in this section engage muscles in the lower body:
the quadriceps, hamstrings, calves, glutes, adductors, and abductors.
Many of the main exercises include variations and modifications;
they all instruct you how to execute each movement to maximize
effectiveness and minimize risk of injury.

SQUAT

This exercise works to strengthen all the largest muscles of the legs and butt, including those hard-to-reach areas such as the quadriceps, glutes, and hamstrings. It improves your lower body mobility and keeps your bones and joints healthy. It also targets your core muscles.

THE BIG **PICTURE**

A squat is a "complex" movement because of the high amount of muscle activation from hips to knees to feet. Make sure that the knees don't go past the toes—incorrect form can lead to knee and lower back injuries. Avoid letting the knees bend inward, hunching the back, lifting the heels off the ground, or starting from the knees. Start with 4 sets of 8–10 reps; discover variations on pp.96–97.

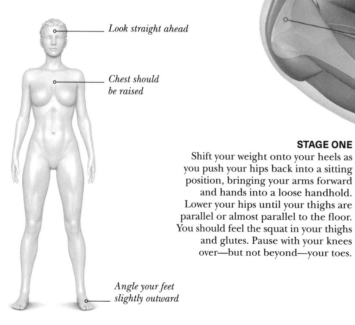

Look straight ahead

Chest should be raised

Angle your feet slightly outward

PREPARATORY STAGE
Start standing with your feet a little wider than hip-width apart, turned out slightly. Hold your body weight mostly in the heels of the feet.

STAGE ONE
Shift your weight onto your heels as you push your hips back into a sitting position, bringing your arms forward and hands into a loose handhold. Lower your hips until your thighs are parallel or almost parallel to the floor. You should feel the squat in your thighs and glutes. Pause with your knees over—but not beyond—your toes.

ANTERIOR-LATERAL VIEW

Upper **body**

The **abdominal** muscles—the **rectus abdominis**, **transverse abdominis**, and **serratus anterior**—should be engaged the whole time. Activating the abdominal muscles helps brace your back, holding the spine in a neutral position. Maintain spinal tension when lowering the body.

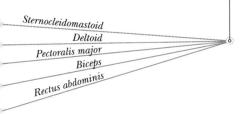

Sternocleidomastoid
Deltoid
Pectoralis major
Biceps
Rectus abdominis

Lower **body**

The **quadriceps** and **adductors** are the primary muscle movers, while the **hamstrings** and **calf** muscles help stabilize the pelvis and knee. Lowering the body down into the squat position is an eccentric movement. Maintain proper form, as the squat can place a lot of tension on the lower body joints.

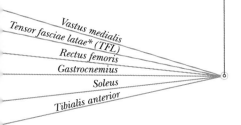

Tensor fasciae latae* (TFL)
Vastus medialis
Rectus femoris
Gastrocnemius
Soleus
Tibialis anterior

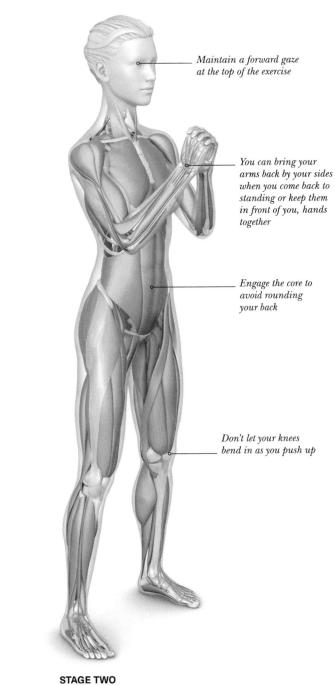

Maintain a forward gaze at the top of the exercise

You can bring your arms back by your sides when you come back to standing or keep them in front of you, hands together

Engage the core to avoid rounding your back

Don't let your knees bend in as you push up

KEY

•-- *Joints*

○— *Muscles*

● Shortening with tension

● Lengthening with tension

● Lengthening without tension

● Held muscles without motion

STAGE TWO

Exhale, tighten the core, and push your feet against the floor and back up to the starting position, chest raised and neck and head both in alignment with spine. Avoid the knees coming inward while pushing up.

» VARIATIONS

These squat variations target different areas of the legs. The chair squat primarily uses all the muscles of the quadriceps, hamstrings, and glutes. The sumo squat with fly recruits the glutes (medius and maximus), as well as the hips, adductors, quadriceps, hamstrings, and calves. The goblet squat targets all the lower body muscles.

66 99

*The knees can **curve** inward in a squat if you have weak glutes or tight hips.*

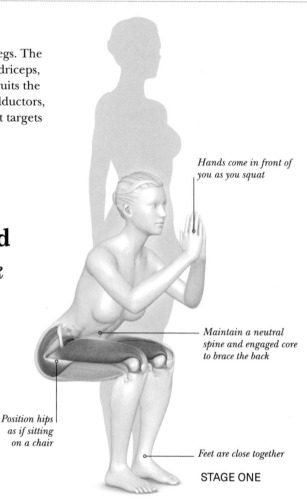

Hands come in front of you as you squat

Maintain a neutral spine and engaged core to brace the back

Position hips as if sitting on a chair

Feet are close together

STAGE ONE

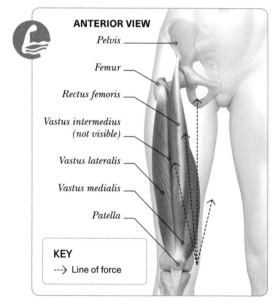

ANTERIOR VIEW

Pelvis

Femur

Rectus femoris

Vastus intermedius (not visible)

Vastus lateralis

Vastus medialis

Patella

KEY

--> Line of force

The quadriceps explained

The quadriceps is a group of four muscles at the front of the thigh, each with different lines of force. The vastus lateralis runs down the outside of your thigh, connecting your femur to your patella, or kneecap. The vastus medialis runs along the inner part of your thigh, between the femur and kneecap. The vastus intermedius, the deepest muscle, lies between the other two vastus muscles. The rectus femoris runs from the hip bone to the kneecap.

CHAIR SQUAT

This exercise strengthens the quads, gluteus maximus, and adductor magnus and builds muscular strength in the hamstrings, abdominals, and obliques. The spinal erector muscles keep the spine extended in the narrow squat.

PREPARATORY STAGE
Start by standing up straight, with your feet less than shoulder-width apart (unlike in the main squat, where your feet are wider). Relax your arms. Inhale.

STAGE ONE
Squat as if you are sitting on a chair: slowly break at the hips and knees to squat as low as you can, dropping your hip lower than parallel to the floor.

STAGE TWO
Slowly stand up by pulling in the hips and straightening the knees. Exhale to pull yourself up until your knees and hips are fully locked out.

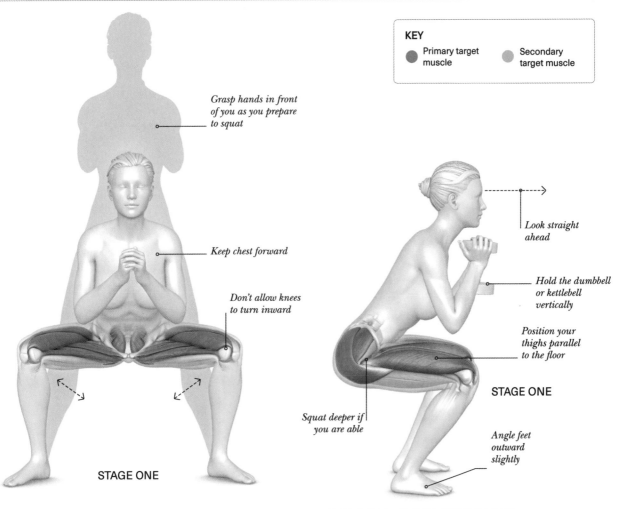

KEY

● Primary target muscle
● Secondary target muscle

Grasp hands in front of you as you prepare to squat

Keep chest forward

Don't allow knees to turn inward

STAGE ONE

Look straight ahead

Hold the dumbbell or kettlebell vertically

Position your thighs parallel to the floor

STAGE ONE

Squat deeper if you are able

Angle feet outward slightly

SUMO SQUAT AND SUMO FLY

This exercise strengthens the glutes, quads, hamstrings, hip flexors, calves, and core. The "flapping" motion of the thighs in the sumo fly places greater emphasis on the hips, as well as the adductor muscles of the inner thighs.

PREPARATORY STAGE
Stand with your feet wide and turn your toes out to 45 degrees. Make sure your head and neck are in a neutral position, spine in alignment, and weight evenly balanced.

STAGE ONE
Start bending at the hips and knees, slowly pushing the butt back. Lower your body down until your thighs are parallel to the floor, then "flap" your knees in and out.

STAGE TWO
Stand back up to the starting position, making sure the knees don't turn inward. Also keep the spine and neck in a neutral position.

DUMBBELL GOBLET SQUAT

The goblet squat is a total body exercise: it works all the major muscle groups of the lower body, including the quads, glutes, and hamstrings. Holding the weight in front of the body engages the quads in particular.

PREPARATORY STAGE
Stand with your feet slightly wider than hip-distance apart, your toes angled slightly outward. Hold your dumbbell in both hands, gripping its head as if cupping a goblet.

STAGE ONE
Inhale, press your hips back and begin bending your knees to perform the squat, keeping your chest upright as you lower. Distribute your weight evenly between both feet.

STAGE TWO
Exhale, press through your heels, and drive out of the squat to return to the starting position. Press your hips forward at the top of the squat to engage your glutes.

RIGHT AND LEFT
SPLIT SQUAT

The split squat is a unilateral leg exercise—working one leg at a time. It strengthens your lower body and improves balance, stability, and hip mobility. The hamstrings, quadriceps, glutes, and core are all strengthened, too. The hamstrings offer balance, stability, and strength in the lowering phase of the split, which can boost the strength and size of these muscles.

THE **BIG PICTURE**

When you are focusing on one leg at a time, it's easier to strengthen that leg and avoid injury. Stance placement is important here; make sure the legs aren't too close or too far apart. Keep your shoulders back, not hunched over. Always make sure the core is engaged and that the front knee doesn't go beyond the toe. If you are a beginner, start with 1 set of 8–12 reps, then repeat for the other leg. Build up to 3 sets as you become used to the exercise. Holding a dumbbell in each hand while doing the split squat increases the challenge of this exercise.

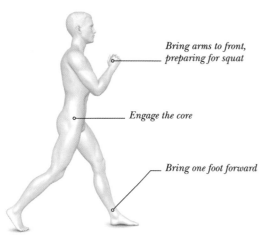

Bring arms to front, preparing for squat

Engage the core

Bring one foot forward

PREPARATORY STAGE
Stand tall with your feet under your hips (shoulder-width apart), toes facing forward. Hold your hands lightly grasped in front of your chest. Take a step forward with one leg. Once you are comfortable and well balanced in the stance, engage the core.

Upper **body**
The **abdominals** are the primary activator in the upper body. The **obliques** and **rectus abdominis** stabilize the core and brace the spine to allow the hips to function in the squat, while the **core** works to resist rotational forces caused by inadequate balance and stability. Using dumbbells creates tension in the **arms**.

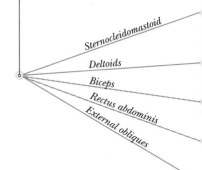

Sternocleidomastoid

Deltoids

Biceps

Rectus abdominis

External obliques

Lower **body**
The **glutes** are responsible for hip extension and stabilizing the pelvis during the split position. The **quadriceps** muscles extend the knee, working together to initiate the split squat. The **hamstrings** engage to balance, stabilize, and aid in the lowering phase of the exercise. The **calf muscles** are also activated.

Tensor fasciae latae

Vastus medialis

Semitendinosus

Gastrocnemius

Rectus femoris

Peroneus longus

Extensor digitorum longus

STAGE ONE
Inhale and lowly squat down, keeping your chest slightly raised, until your back knee almost touches the floor. Your back heel should lift off the floor to balance on the toes. Go as deep as you can, maintaining core engagement throughout the movement. Keep the front knee at a 90-degree angle. Hold for a couple of seconds here.

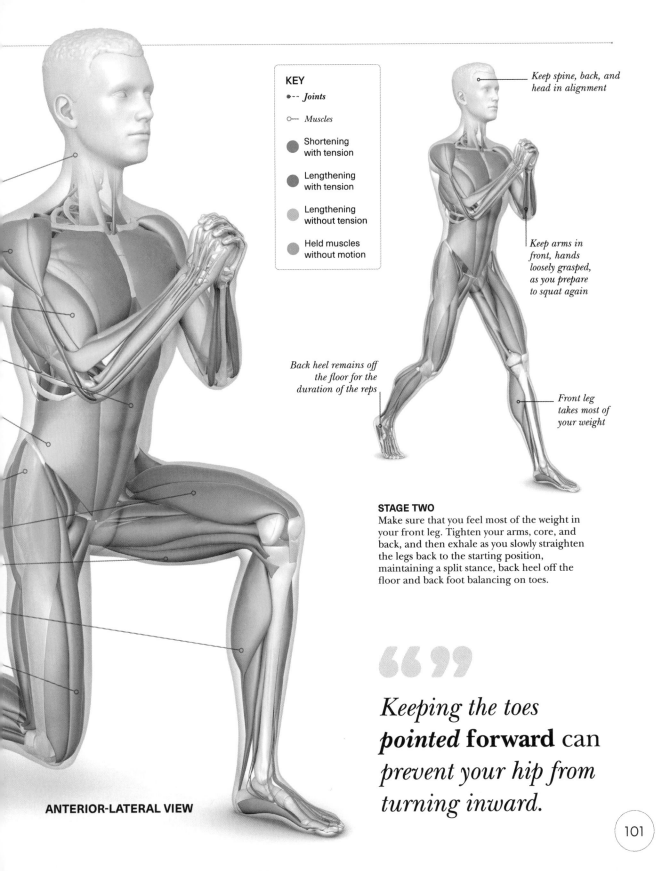

KEY

●-- *Joints*

○— *Muscles*

● Shortening with tension

● Lengthening with tension

● Lengthening without tension

● Held muscles without motion

Keep spine, back, and head in alignment

Keep arms in front, hands loosely grasped, as you prepare to squat again

Back heel remains off the floor for the duration of the reps

Front leg takes most of your weight

STAGE TWO
Make sure that you feel most of the weight in your front leg. Tighten your arms, core, and back, and then exhale as you slowly straighten the legs back to the starting position, maintaining a split stance, back heel off the floor and back foot balancing on toes.

"

Keeping the toes **pointed forward** *can prevent your hip from turning inward.*

ANTERIOR-LATERAL VIEW

» VARIATIONS

Split squats primarily engage all the glute muscles, as well as the abdominals. In the curtsy squat variation, the quads and glutes are isolated. The squat with alternating kickback engages all the muscles used in a squat but further isolates the glutes when you kick each leg back.

" "

*In the down position, don't let your **front knee extend** over your front ankle, which can strain the knee and quads.*

ALTERNATING
CURTSY **SQUAT**

The curtsy squat strengthens the quads and glutes. When your leg crosses back and around, the gluteus medius on the stationary leg fires up. The hip abductors, which bring your thighs together, are also engaged. The calves are activated at the bottom of each lunge as you push back upward.

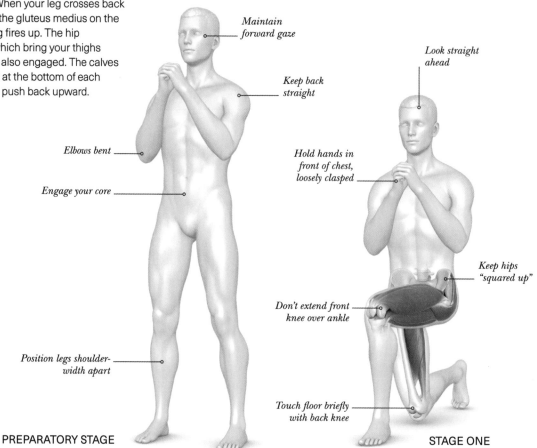

Maintain forward gaze

Look straight ahead

Keep back straight

Elbows bent

Hold hands in front of chest, loosely clasped

Engage your core

Keep hips "squared up"

Don't extend front knee over ankle

Position legs shoulder-width apart

Touch floor briefly with back knee

PREPARATORY STAGE

STAGE ONE

PREPARATORY STAGE
Start in a standing position with your chest held high and your back straight. Your feet should be shoulder-width apart, arms in front of you, and hands facing the chest.

STAGE ONE
Keeping your left foot straight in front of you, step to the back and to the left with your right foot, behind your left leg. Both of your knees will bend and your legs will be "crossed."

STAGE TWO
Tighten the core and drive upward with control back to the standing position. Repeat with the right foot out in front, "curtsying" with the left leg. Continue to alternate squats.

SQUAT AND ALTERNATING KICKBACK

This movement, which combines a squat and a kickback of one leg at a time, engages the hamstrings and the glutes. Repeat the squat and kickback sequence for 30–60 seconds.

KEY

● Primary target muscle

● Secondary target muscle

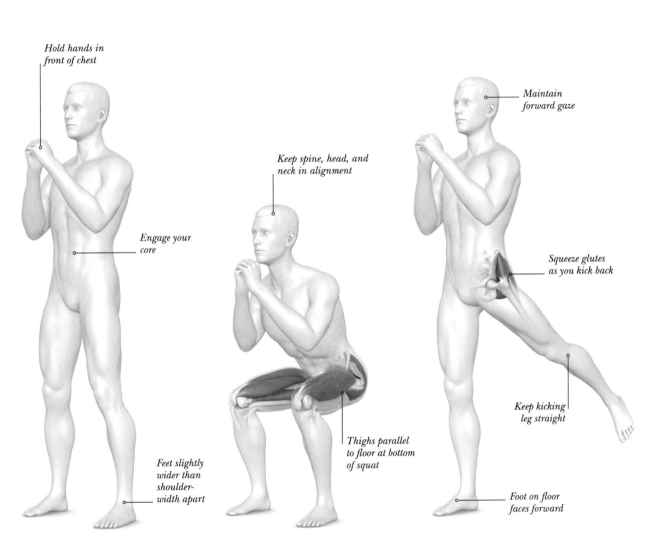

Hold hands in front of chest

Engage your core

Feet slightly wider than shoulder-width apart

Keep spine, head, and neck in alignment

Thighs parallel to floor at bottom of squat

Maintain forward gaze

Squeeze glutes as you kick back

Keep kicking leg straight

Foot on floor faces forward

PREPARATORY STAGE
Start with your feet slightly wider than shoulder-width apart, arms in front of you and hands loosely grasped together in front of your upper chest.

STAGE ONE
Tighten the core, inhale, and slowly bend the knees into a squat, keeping the chest slightly raised and the spine, neck, and head in alignment. Don't let the knees go past the toes.

STAGE TWO
Raise up to standing and exhale as you shift your weight to the right leg. Kick the left leg behind you, pause for 2 seconds, then return to the squat, alternating the kickbacks.

CRAB WALK

This lateral "crab" walk fully engages your glutes and hip abductors and also strengthens all major muscles in your hips, thighs, and legs. It improves your flexibility and stability and helps prevent injuries, and it is particularly helpful for anyone doing sports that require running, jumping, and twisting.

THE BIG PICTURE

The half squat—the starting position and the position for the whole exercise—is halfway between a full squat and a standing position. If you're doing it correctly, you should feel it in your gluteus medius. In your partial squat position, keep your knees bent and aligned over the middle of your foot. This ensures you are targeting the right muscles and not straining your knees. If just beginning, start by performing this movement for 30 seconds. Slowly work up to 60 seconds. Repeat for 3–4 sets, making sure you take the same amount of steps with each foot.

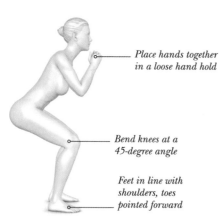

Place hands together in a loose hand hold

Bend knees at a 45-degree angle

Feet in line with shoulders, toes pointed forward

PREPARATORY STAGE
Stand with your feet shoulder-width apart. Bend your knees slightly and go down into a half-squat position to activate the gluteus medius. Distribute weight evenly over both feet. Engage the core and raise your chest slightly.

POSTERIOR-LATERAL VIEW

KEY

•-- *Joints*

o-- *Muscles*

● Shortening with tension

● Lengthening with tension

● Lengthening without tension

● Held muscles without motion

Lower **body**

All the **quadriceps** muscles are engaged to extend the knee. The **abductors** work with the **gluteus medius** and **maximus** to lift your legs laterally. They include the **tensor fasciae latae, superior** and **inferior gemellus,** and **piriformis**.

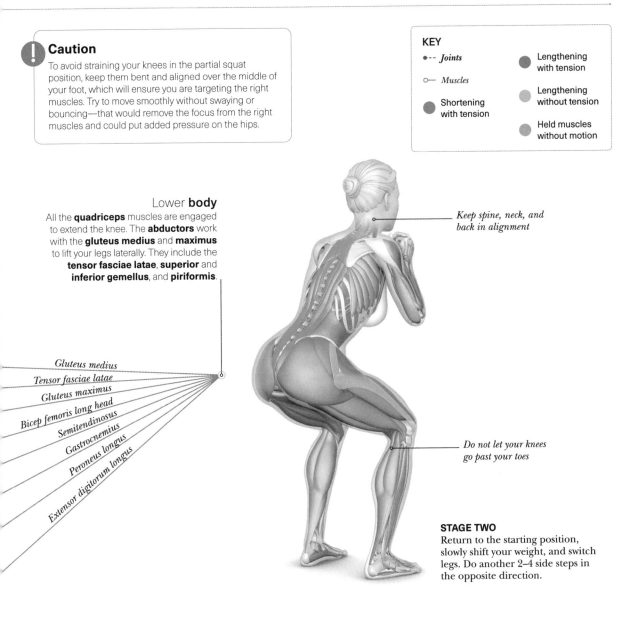

Keep spine, neck, and back in alignment

Gluteus medius
Tensor fasciae latae
Gluteus maximus
Bicep femoris long head
Semitendinosus
Gastrocnemius
Peroneus longus
Extensor digitorum longus

Do not let your knees go past your toes

STAGE TWO
Return to the starting position, slowly shift your weight, and switch legs. Do another 2–4 side steps in the opposite direction.

STAGE ONE
Once in the half-squat position, step one leg to the right and follow it by taking a step sideways in the same direction with the other leg. Continue this sideways movement for 2–4 reps, keeping your hips level. Maintain a low, forward-facing posture and a straight back.

66 99

*A strong gluteus medius **stabilizes the** **hip** and reduces lateral stress on the knees.*

ALTERNATING SNATCH

Also known as the single-arm dumbbell snatch,
this is a powerful compound movement that works the whole body. It improves your speed and agility while strengthening the quadriceps, hamstrings, and glutes. It also targets your back and shoulder muscles.

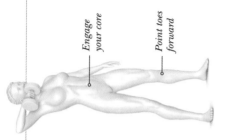

Engage your core

Point toes forward

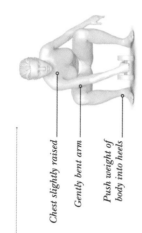

Chest slightly raised

Gently bent arm

Push weight of body into heels

THE BIG PICTURE

In addition to activating your muscles, the dumbbell or kettlebell snatch may help increase cardiorespiratory fitness. When lowering your body to transfer the weight to the other hand, hinge at your hips and bend your knees without rounding your back or looking down. Focus on using the momentum from your lower body to transfer the weight rather than relying on your shoulder and arms. If beginning, start with 1 set of 8–12 reps.

PREPARATORY STAGE

Stand with feet shoulder-width apart, a dumbbell on the floor between your feet. Bend the knees, hinge your hips, and lower into a squat position. Grab the dumbbell, rotating your elbow and shoulder externally before rising.

STAGE ONE

Keep your shoulders back, chest slightly raised, and eyes facing straight ahead. With your weight in your heels, stand up explosively, lifting the weight toward your right shoulder.

Upper **body** and **core**

The **latissimus dorsi** muscles in the back engage to lift the dumbbell from the floor, while the **spinal erectors** maintain stability of the spine during the hip extension through the top of the movement. The **rotator cuff** and **deltoid** shoulder muscles assist in raising the dumbbell overhead. Your **core** muscles are active throughout to stabilize the body.

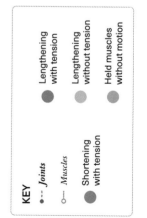

Biceps

Triceps

Latissimus dorsi

Pectoralis major

External obliques

Rectus abdominis

KEY

- ● - - *Joints*
- ○ — *Muscles*
- ● Shortening with tension
- ● Lengthening with tension
- ● Lengthening without tension
- ● Held muscles without motion

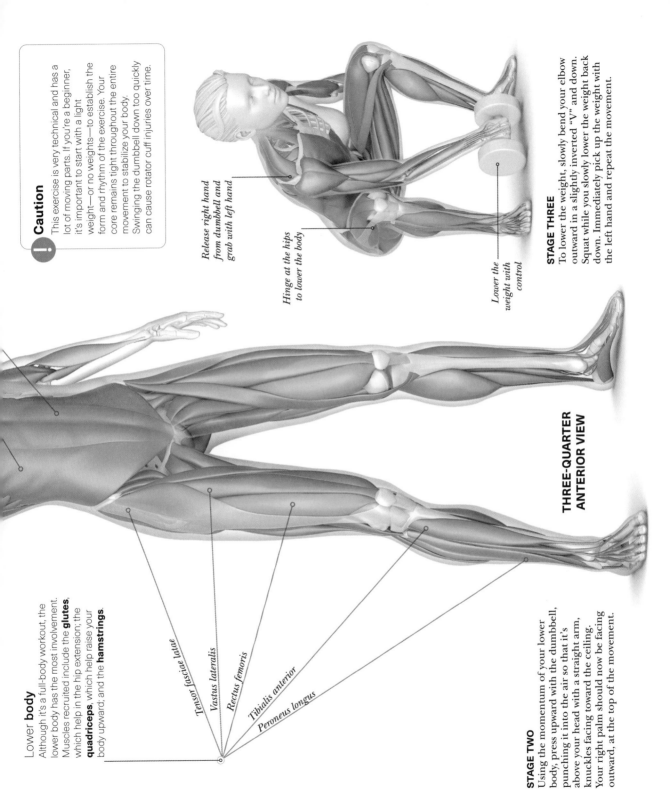

Caution

This exercise is very technical and has a lot of moving parts. If you're a beginner, it's important to start with a light weight—or no weights—to establish the form and rhythm of the exercise. Your core remains tight throughout the entire movement to stabilize your body. Swinging the dumbbell down too quickly can cause rotator cuff injuries over time.

Release right hand from dumbbell and grab with left hand

Hinge at the hips to lower the body

Lower the weight with control

STAGE THREE
To lower the weight, slowly bend your elbow outward in a slightly inverted "V" and down. Squat while you slowly lower the weight back down. Immediately pick up the weight with the left hand and repeat the movement.

THREE-QUARTER ANTERIOR VIEW

Lower **body**
Although it's a full-body workout, the lower body has the most involvement. Muscles recruited include the **glutes**, which help in the hip extension; the **quadriceps**, which help raise your body upward; and the **hamstrings**.

Tensor fasciae latae

Vastus lateralis

Rectus femoris

Tibialis anterior

Peroneus longus

STAGE TWO
Using the momentum of your lower body, press upward with the dumbbell, punching it into the air so that it's above your head with a straight arm, knuckles facing toward the ceiling. Your right palm should now be facing outward, at the top of the movement.

107

ALTERNATING
LATERAL LUNGE

This exercise improves balance, stability, and strength. The step out to the side engages your muscles differently from other lunges: it works your inner and outer thighs in addition to targeting your quadriceps, hips, and legs. The lateral lunge also engages the glutes and can improve athletic performance and agility.

THE **BIG PICTURE**

Avoid rounding the back, and don't bring the body forward too much. The knee placement is important, as the knee shouldn't go past the toes. To elevate this workout, everything will remain the same except you add weight to increase the resistance of the workout. You can do this progression with a barbell on the back or a dumbbell held in each hand.

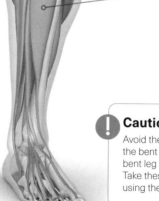

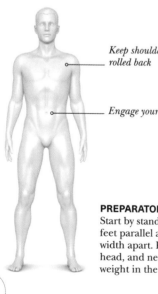

Keep shoulders rolled back

Engage your core

PREPARATORY STAGE
Start by standing tall with your feet parallel and shoulder-width apart. Keep your spine, head, and neck neutral, your weight in the heels of the foot.

! Caution
Avoid the knees passing the toes on the bent leg. Make sure the heel of the bent leg remains planted on the floor. Take these lunges as deep as possible, using the full range of motion.

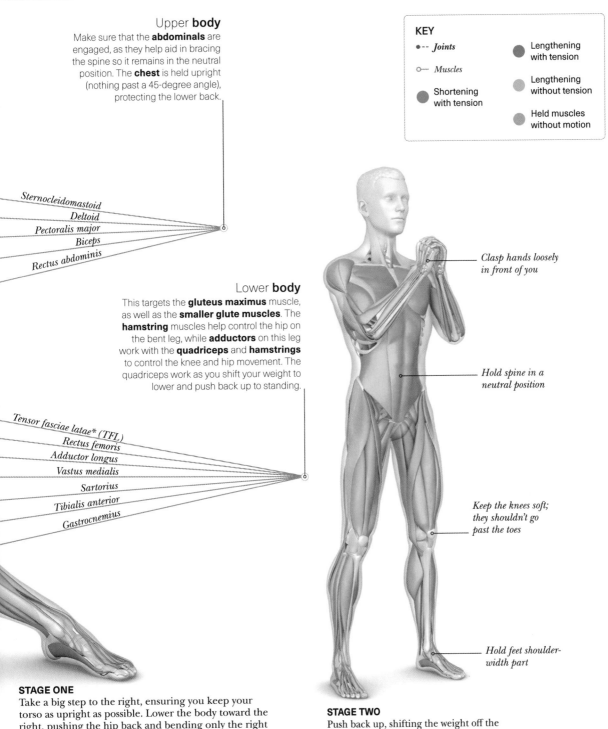

Upper **body**

Make sure that the **abdominals** are engaged, as they help aid in bracing the spine so it remains in the neutral position. The **chest** is held upright (nothing past a 45-degree angle), protecting the lower back.

Sternocleidomastoid
Deltoid
Pectoralis major
Biceps
Rectus abdominis

Lower **body**

This targets the **gluteus maximus** muscle, as well as the **smaller glute muscles**. The **hamstring** muscles help control the hip on the bent leg, while **adductors** on this leg work with the **quadriceps** and **hamstrings** to control the knee and hip movement. The quadriceps work as you shift your weight to lower and push back up to standing.

Tensor fasciae latae* (TFL)
Rectus femoris
Adductor longus
Vastus medialis
Sartorius
Tibialis anterior
Gastrocnemius

KEY

- •-- *Joints*
- ○— *Muscles*
- Shortening with tension
- Lengthening with tension
- Lengthening without tension
- Held muscles without motion

Clasp hands loosely in front of you

Hold spine in a neutral position

Keep the knees soft; they shouldn't go past the toes

Hold feet shoulder-width part

STAGE ONE

Take a big step to the right, ensuring you keep your torso as upright as possible. Lower the body toward the right, pushing the hip back and bending only the right knee. Descend until the knee is bent at around 90 degrees, keeping your opposite leg straight out to the other side. As you bend, bring your arms to the front.

STAGE TWO

Push back up, shifting the weight off the right leg and back to the center. Return to the starting position and repeat the exercise on the left side, keeping the same form.

» VARIATIONS

Because of the angle, lateral (or side) lunges specifically target the hip adductors, used to bring the thighs together, and the hip abductors, used to bring the legs away from the body and rotate the leg at the hip. These variations activate the same muscle groups, adding more challenge to the original exercise.

Neck and head are in alignment with the spine

Keep torso straight

Hold your arms in front of you as you perform the lunge

Bend the left knee at 90-degree angle

Flex the toes

STAGE ONE

Keep looking straight ahead

Keep your chest upright

Engage your core

Right thigh is parallel to the floor

Keep hip "squared" to the torso

Flex toes of back leg

Foot points forward

STAGE ONE

ALTERNATING BACK LUNGE

Back lunges strengthen the quads at the front of your upper legs. As you progress, you can add some weights to the move, making sure you don't choose weights that are too heavy for your fitness level.

PREPARATORY STAGE
Start by standing straight with your feet shoulder-width apart and your toes pointed forward. Engage the core and hold hands in front of your chest in a loose grasp.

STAGE ONE
Slowly move your right leg back behind your body as if to kneel, but keep your knee just off the floor. At the same time, bend your left knee and lower your hips. Keep your torso straight and stop when your knee is at a 90-degree angle and your left thigh is parallel to the floor, making sure the knee doesn't go past the toes. Pause, then push with your left leg, squeezing your glutes to stand up, while simultaneously returning your right leg to its starting position. Repeat on the left leg.

FRONT CURTSY LUNGE

The curtsy lunge is great for building lower body strength and stability: it targets the quads, glutes, hip abductors, and inner thighs. The gluteus medius, often underactive, makes the curtsy lunge an important move.

PREPARATORY STAGE
Stand with your feet shoulder-width apart and your arms in front of you, hands held together in front of your upper chest.

STAGE ONE
Put your weight onto your left foot, step your right foot forward, and in front of your left knee to do a forward "curtsy" movement, keeping the spine in alignment. Stop when your right thigh is parallel to the floor.

STAGE TWO
Begin to straighten your right leg, pushing up through your heel, returning your right and left foot at the same time to the starting position. Repeat on the other side.

WALKING LUNGE
WITH DUMBBELLS

This walking lunge variation adds challenge and coordination to the stationary lunge. Begin using just body weight until you can remain balanced and coordinated as you get used to this movement. Once you are confident, you can add in some dumbbells.

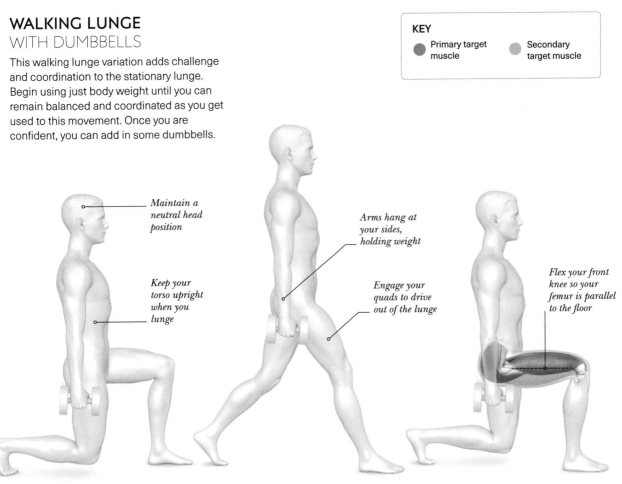

Maintain a neutral head position

Keep your torso upright when you lunge

Arms hang at your sides, holding weight

Engage your quads to drive out of the lunge

Flex your front knee so your femur is parallel to the floor

PREPARATORY STAGE
Begin standing with your feet shoulder-width apart. Breathe in and stride forward into the lunge position, bending your back knee so that it is just off the floor.

STAGE ONE
Exhale to power upright from the lunge and immediately take a stride forward with your other leg. Stand tall and keep your abs engaged throughout.

STAGE TWO
Breathe in as you drive your hip down and your front knee forward, allowing your back knee to flex, as before. Repeat, alternating legs as you move along.

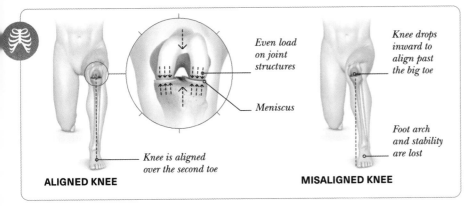

Even load on joint structures

Meniscus

Knee is aligned over the second toe

ALIGNED KNEE

Knee drops inward to align past the big toe

Foot arch and stability are lost

MISALIGNED KNEE

Knee **alignment**
When performing lunges, the knee should be over the foot with the kneecap in line with the outside two toes. Both knees should be at a 90-degree angle. A very common misalignment is for the knee to fall inward, toward the midline of the body, known as valgus collapse. This collapse puts uneven stress on the joint, which over time can lead to pain and injuries.

CALF RAISE

The standing calf raise targets the calf muscle, specifically the gastrocnemius. Movements that are done with a bent knee (flexion) target the soleus, which attaches below the knee joint. Strong, flexible calf muscles will help maintain healthy knees and also promote ankle strength.

THE BIG PICTURE

The low-impact movement of the calf raise is a perfect exercise for beginners. You can perform it using a calf-raise machine or standing in front of a wall using your fingertips for balance. With both, make sure you balance on the balls of your feet while you raise yourself up to your tiptoes. Keep the knees soft, the feet shoulder-width apart, and the toes parallel. If you're new to weight lifting, start with a set of light weights and plan to do 10–12 repetitions for 3 sets.

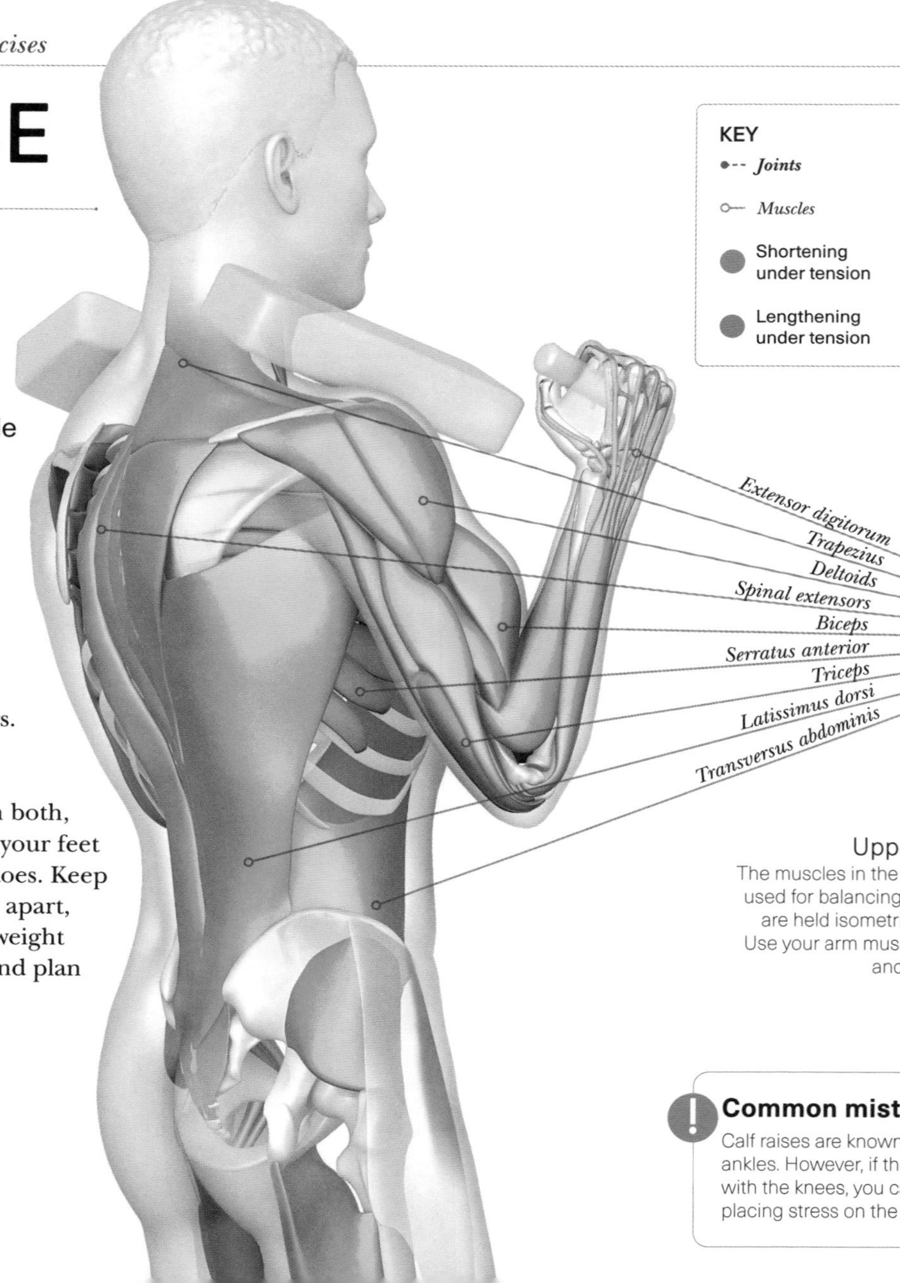

KEY

-- Joints

○— Muscles

● Shortening under tension

● Lengthening under tension

● Lengthening without tension (stretching)

● Held muscles without motion

Extensor digitorum

Trapezius

Deltoids

Spinal extensors

Biceps

Serratus anterior

Triceps

Latissimus dorsi

Transversus abdominis

Upper body and arms

The muscles in the **upper body** are mostly used for balancing, while the **abdominals** are held isometrically to brace the spine. Use your arm muscles to hold the handles and help stabilize the body.

! Common mistakes

Calf raises are known to strengthen the ankles. However, if they aren't kept in line with the knees, you can run the risk of placing stress on the Achilles tendon.

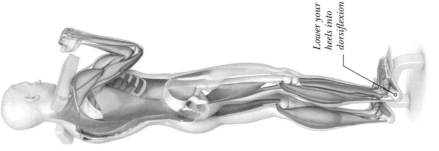

Lower your heels into dorsiflexion

Lower **body**

Standing calf raises activate the **gastrocnemius** and **soleus**. These muscles assist in ankle flexion and extension, propelling running and jumping. The gastrocnemius also works in tandem with **hamstrings** to control knee flexion, while the soleus maintains balance.

Adductor magnus

Knee

Tibialis anterior

Gastrocnemius

Soleus

Peroneus longus

Ankle

Extensor digitorum longus

Flexor hallucis longus

Abductor digiti minimi

POSTERIOR-LATERAL VIEW

STAGE ONE

Inhale to engage the core. While exhaling, slowly raise your heels until you are on tiptoes, keeping your knees extended (but not locked). Pause for 1–2 seconds at the top of the raise.

STAGE TWO

Breathe in and slowly lower your heels all the way back down, pausing for 1–2 seconds at the bottom. Adjust your form and repeat stages 1 and 2.

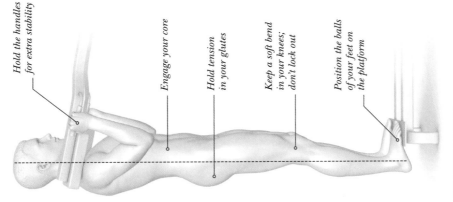

Hold the handles for extra stability

Engage your core

Hold tension in your glutes

Keep a soft bend in your knees; don't lock out

Position the balls of your feet on the platform

PREPARATORY STAGE

Set the weights to your current fitness level. Place your shoulders under the pad and stand with the balls of your feet on the edge of the step, feet shoulder-width apart and parallel. Tighten your core for balance. Lower the heels slowly into the starting position.

STEP-UP WITH DUMBBELLS

This exercise strengthens the quadriceps and the posterior chain, as well as challenging the muscles of the core. It's a great exercise at any level.

THE **BIG PICTURE**

The front leg (on the step) is the driver in this exercise—don't push off the floor with your back foot. Beginners should start with body weight step-ups on a lower step, moving on to a 12-in (30-cm) step and weights when more confident. It's important to make sure the whole foot is firmly planted on the step prior to starting the movement. Plan to do 10–12 repetitions for 3 sets if just beginning. You can alternate the legs each rep, or perform 10 reps on one side, then 10 reps on the other side.

Hips and legs

Your **quadriceps** are the main muscles activated during the step-up, while your **hamstrings** work to stabilize your knee joint and your torso at the hip. Your **glutes** engage to pull your thigh back into line with your torso and help keep your torso upright. The **gluteus maximus** and the smaller side hip muscle of the **gluteus medius** are highly stimulated during the move.

Gluteus medius
Gluteus fascia latae
Tensor fascia latae
Iliopsoas

Upper **body** and **abdominals**

Your **abdominals** engage to brace the back to keep you from leaning forward or backward. Your **internal and external obliques** contract to keep you from wobbling from side to side as you step up and down.

Sternocleidomastoid
Trapezius
Deltoids
Pectoralis minor
Biceps
Triceps
Brachialis
Spine
Rectus abdominis
Brachioradialis
Transversus abdominis

ANTERIOR-LATERAL VIEW

KEY
● -- *Joints*
○— *Muscles*
● Shortening with tension
● Lengthening with tension
● Lengthening without tension
● Held muscles without motion

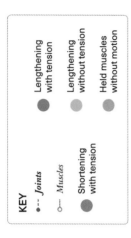

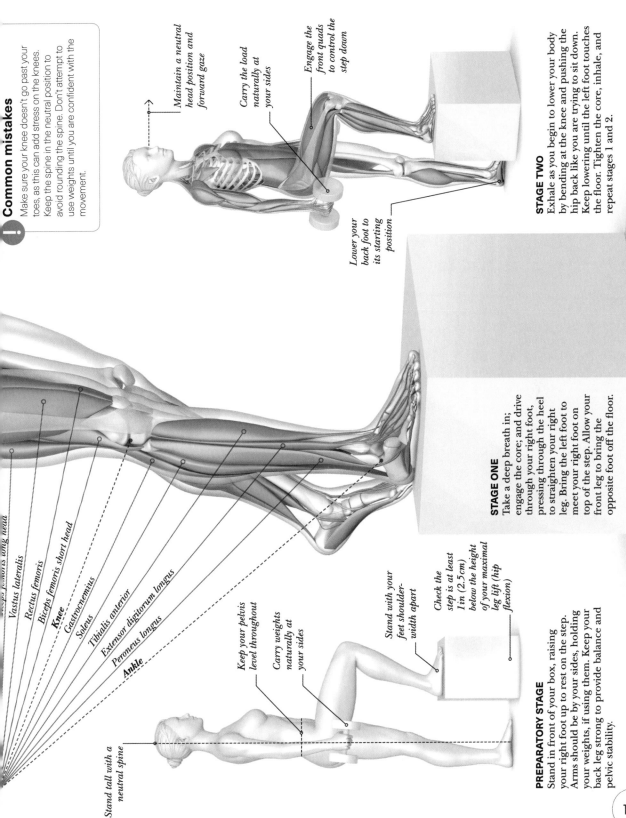

! Common mistakes

Make sure your knee doesn't go past your toes, as this can add stress on the knees. Keep the spine in the neutral position to avoid rounding the spine. Don't attempt to use weights until you are confident with the movement.

Maintain a neutral head position and forward gaze

Carry the load naturally at your sides

Engage the front quads to control the step down

Lower your back foot to its starting position

STAGE TWO

Exhale as you begin to lower your body by bending at the knee and pushing the hip back like you are trying to sit down. Keep lowering until the left foot touches the floor. Tighten the core, inhale, and repeat stages 1 and 2.

STAGE ONE

Take a deep breath in; engage the core; and drive through your right foot, pressing through the heel to straighten your right leg. Bring the left foot to meet your right foot on top of the step. Allow your front leg to bring the opposite foot off the floor.

Biceps femoris long head
Vastus lateralis
Rectus femoris
Biceps femoris short head
Knee
Gastrocnemius
Soleus
Tibialis anterior
Extensor digitorum longus
Peroneus longus
Ankle

Keep your pelvis level throughout

Carry weights naturally at your sides

Stand with your feet shoulder-width apart

Check the step is at least 1in (2.5cm) below the height of your maximal leg lift (hip flexion)

PREPARATORY STAGE

Stand in front of your box, raising your right foot up to rest on the step. Arms should be by your sides, holding your weights, if using them. Keep your back leg strong to provide balance and pelvic stability.

Stand tall with a neutral spine

115

ALTERNATING
TOE TAP

This cardiovascular exercise strengthens the quadriceps, hamstrings, calves, glutes, and hip flexors. It also requires engagement from the core muscles. The toe tap enhances speed, agility, endurance, and overall athletic performance.

THE **BIG PICTURE**

Start out with a platform, step, or box that's the appropriate height for your current fitness level. Engaging your core will help with balance, stability, and support. It also helps bring the knees up quicker with more power. Start with 30 seconds of reps, progress to 45 seconds, then go up to 60 seconds with a slightly higher platform.

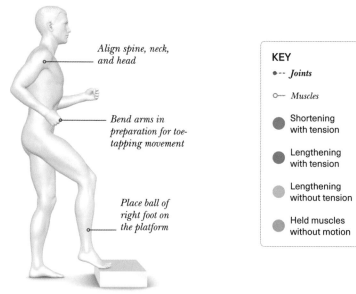

Align spine, neck, and head

Bend arms in preparation for toe-tapping movement

Place ball of right foot on the platform

KEY

●-- *Joints*

○— *Muscles*

● Shortening with tension

● Lengthening with tension

○ Lengthening without tension

● Held muscles without motion

PREPARATORY STAGE
Set your feet shoulder-width apart and have your arms at your sides. Raise your right foot and place the ball of your foot on the platform. Your left foot will remain flat on the floor, your arms bent at a 45–90 degree angle in preparation for the toe-tapping movement.

ANTERIOR-LATERAL VIEW

Abdominals and cardiovascular endurance

The alternating toe tap provides you with a fantastic cardio workout, increasing your cardiovascular endurance. The **abdominals**, particularly the **rectus abdominis**, work to stabilize and hold the body upright by bracing the spine and keeping it in neutral.

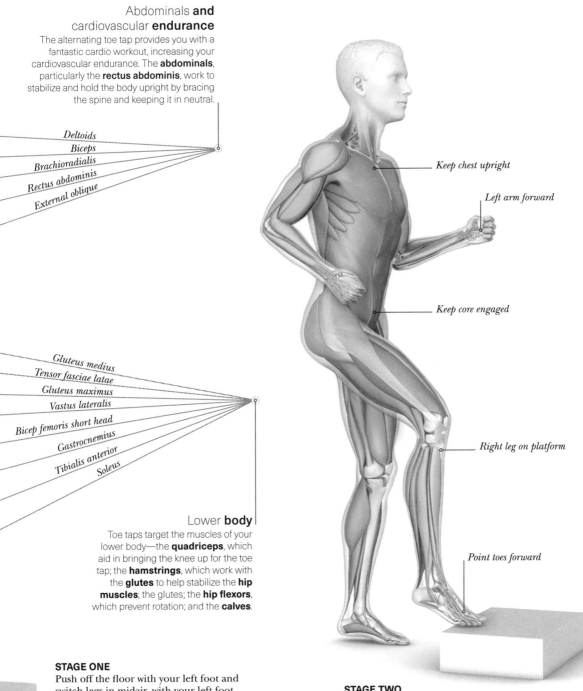

Deltoids
Biceps
Brachioradialis
Rectus abdominis
External oblique

Keep chest upright

Left arm forward

Keep core engaged

Gluteus medius
Tensor fasciae latae
Gluteus maximus
Vastus lateralis
Bicep femoris short head
Gastrocnemius
Tibialis anterior
Soleus

Right leg on platform

Point toes forward

Lower body

Toe taps target the muscles of your lower body—the **quadriceps**, which aid in bringing the knee up for the toe tap; the **hamstrings**, which work with the **glutes** to help stabilize the **hip muscles**; the glutes; the **hip flexors**, which prevent rotation; and the **calves**.

STAGE ONE

Push off the floor with your left foot and switch legs in midair, with your left foot now touching the platform and your right foot on the floor. Keep your arms bent and alternate bringing them forward and back, as if you were running on the spot. This movement should be quite quick.

STAGE TWO

Practice repeating alternating toe taps slowly until you feel comfortable with the movement and your form is correct. By increasing speed, time, and the height of your step, you can increase the amount of calories burned.

117

SINGLE-LEG
DEADLIFT

This unilateral exercise (working one leg at a time) strengthens the glutes—the gluteus maximus, gluteus medius, and gluteus minimus. The glutes are part of the "posterior chain," which also includes the hamstrings and lower back muscles. All of these muscles help maintain an upright posture and balance the body.

Deltoids
Biceps
Serratus anterior
Rectus abdominis
Transversus abdominis

THE **BIG PICTURE**

As you bend forward into the deadlift, your body should be in a straight line without either arching your spine or rounding it. The spine, neck, and head should always remain in alignment. Complete both stages of the exercise slowly and with control. Start with 5–10 reps per leg. As you improve form and strength, you can increase the weight.

KEY

•-- *Joints*

○— *Muscles*

● Shortening with tension

● Lengthening with tension

● Lengthening without tension

● Held muscles without motion

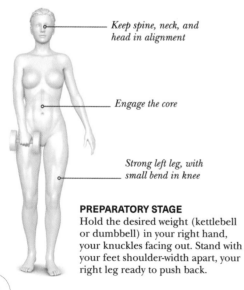

Keep spine, neck, and head in alignment

Engage the core

Strong left leg, with small bend in knee

PREPARATORY STAGE
Hold the desired weight (kettlebell or dumbbell) in your right hand, your knuckles facing out. Stand with your feet shoulder-width apart, your right leg ready to push back.

STAGE ONE
Press into the supporting (left) leg as you slide the nonsupporting (right) leg back with control, allowing your upper body to move forward with your hip as the hinge. If at any point during the exercise you start to lose balance, simply touch the free-floating leg lightly to the floor to regain balance and tighten your core.

Upper **body** and abdominals

The **spinal erectors**, which help support your spine and provide flexibility when bending in multiple directions, are targeted in this exercise. The **trapezius**, forearms, and mid/lower back contract to control the weight. The **abdominals** and **obliques** contract isometrically to stabilize your body and brace the spine, holding it in a neutral position.

! Caution

Rounding the back could lead to injuries and/or back pain. Your back leg should be kept straight, in line with your spine, with your body forming one straight line from the neck to the heel. Bending the back leg takes the spine out of alignment.

> *A unilateral exercise like the **single-leg deadlift** can help reduce injuries and strengthen your* lower **back**.

LATERAL VIEW

Semimembranosus

Adductor longus

Gluteus maximus

Gluteus medius

Tensor fasciae latae

Bicep femoris long head

Vastus lateralis

Biceps femoris short head

Gastrocnemius

Peroneus longus

Push hips forward at top of movement

Weight returns to starting position

Hamstring and glute muscles work to pull you back up

Right leg ready to hinge back

Lower **body**

The primary focus in this exercise is on the posterior chain of the body—the **glutes** and **hamstrings**. The hamstrings provide the muscular force we need to pull, push, and move in the deadlift. Of the glutes, which form the central part of the posterior chain, the **gluteus maximus** muscle is targeted the most.

STAGE TWO

Complete the movement by pulling the weight upward and bringing the elevated leg back to the starting position. Keep coming up until the leg is back on the floor and the planted leg is straight. Complete the reps, then switch legs.

GLUTE BRIDGE

This exercise not only targets the glutes, it strengthens the rectus abdominis, obliques, and quadriceps. In addition, you also target the erector spinae muscle, which runs the length of your back from neck to tailbone. The strong core resulting from the glute bridge will improve your posture and can help ease lower back pain.

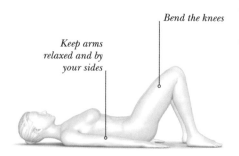

Keep arms relaxed and by your sides

Bend the knees

PREPARATORY STAGE
Start by lying on your back with your hands at your sides, palms on the floor, knees bent, and feet flat on the floor. Tighten your abdominals by pushing your lower back into the ground and squeezing the glute muscles before you push up.

THE **BIG PICTURE**

It's important not to raise the hips too high in this exercise, as putting too much pressure on your lower back can lead to strain. Keeping your abdominals engaged will ensure you don't arch your back excessively. If your hips are dropping as you try to hold the bridge position, lower your pelvis back down to the floor and start again. Beginners can hold the bridge for a few seconds at a time, completing 1 set of 8–12 reps. To progress, increase the sets and the time the bridge is held.

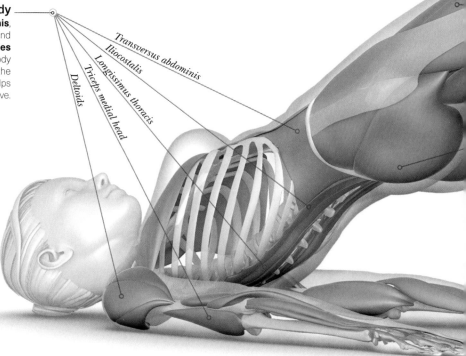

Upper body
The **rectus abdominis**, **transverse abdominis**, and **internal and external obliques** work to help stabilize the body while you move up and down in the bridge. Tightening the core helps brace the spine during the move.

Transversus abdominis

Iliocostalis

Longissimus thoracis

Triceps medial head

Deltoids

STAGE ONE
Exhale and, pushing from your heels, slowly ease your hips up to create a straight diagonal line from your knees to your shoulders. Keep your hands on the floor as your hips raise up into the bridge. Make sure the core is engaged, belly button to spine.

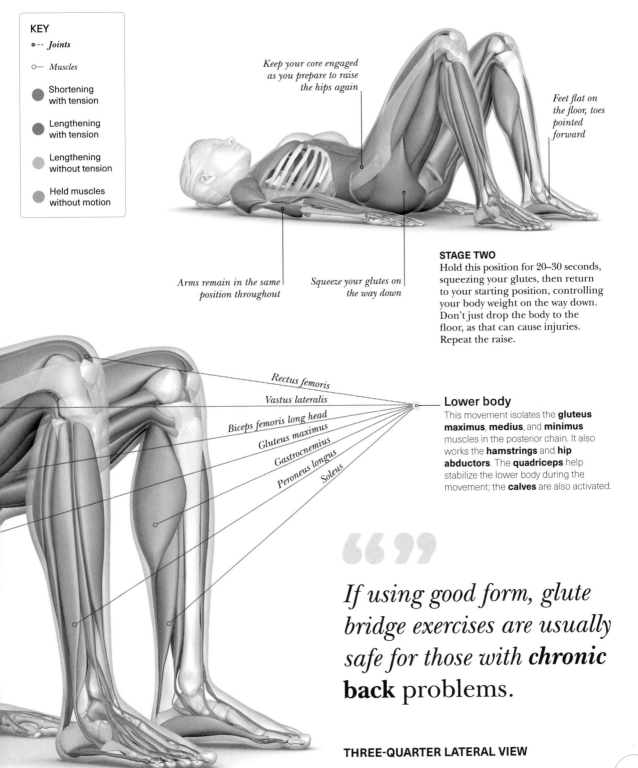

KEY

●--- *Joints*

○— *Muscles*

● Shortening
with tension

● Lengthening
with tension

○ Lengthening
without tension

● Held muscles
without motion

*Keep your core engaged
as you prepare to raise
the hips again*

*Feet flat on
the floor, toes
pointed
forward*

*Arms remain in the same
position throughout*

*Squeeze your glutes on
the way down*

STAGE TWO

Hold this position for 20–30 seconds,
squeezing your glutes, then return
to your starting position, controlling
your body weight on the way down.
Don't just drop the body to the
floor, as that can cause injuries.
Repeat the raise.

Rectus femoris

Vastus lateralis

Biceps femoris long head

Gluteus maximus

Gastrocnemius

Peroneus longus

Soleus

Lower body

This movement isolates the **gluteus
maximus**, **medius**, and **minimus**
muscles in the posterior chain. It also
works the **hamstrings** and **hip
abductors**. The **quadriceps** help
stabilize the lower body during the
movement; the **calves** are also activated.

"

*If using good form, glute
bridge exercises are usually
safe for those with **chronic
back** problems.*

THREE-QUARTER LATERAL VIEW

121

» VARIATIONS

The primary glute bridge exercise engages the glutes, mainly the largest glute muscle: the gluteus maximus. Secondarily, it targets the hamstrings and the transverse abdominis. The variations add challenge to the original exercise while recruiting the same muscles.

*Raising the hips **too high** puts **pressure** on your lower back that can lead to strain.*

Flare the knees outward

Push hips up from floor

Engage your core

Focus on lifting from the heels

Raise leg to a 90-degree angle

STAGE ONE

STAGE ONE

Bring soles of feet together

Keep shoulders on floor

Keep other leg firmly on the floor

Hold head in a neutral position

BUTTERFLY GLUTE BRIDGE

This exercise is absolutely amazing in that it targets all three muscles in the glutes. Because of the rotation in the hips, the butterfly glute bridge provides greater activation than the regular glute bridges.

PREPARATORY STAGE
Start by lying on your back with your hands at your sides, palms on the floor. Bring the soles of your feet together and flare your knees out. Tighten your abs—push your lower back against the floor and squeeze the glute muscles—before you push up.

STAGE ONE
Inhale and slowly push the hip flexors up as you are exhaling. As you lift up, push the hip forward and push the knees out to either side. Hold for a few seconds.

STAGE TWO
Inhale as you lower your hips back down to the floor, squeezing your glutes. Repeat the bridge for 8 reps.

ALTERNATING SINGLE-LEG GLUTE BRIDGE

Because it's a unilateral movement, this variation trains you to balance. It engages the hamstrings, hip flexors, lower back, abdominals, and all three glute muscles.

PREPARATORY STAGE
Begin in the starting position for a basic glute bridge (see pp.120–121). Tighten the abs by pushing your lower back into the floor and squeezing your glutes.

STAGE ONE
Raise your pelvis up into the bridge while raising your left leg. Keep your right leg on the floor as you do this. Holding your position, slowly lower your left leg, heel first, along with your hips, until it's back on the floor.

STAGE TWO
Switch sides, lifting the right leg. When performing this exercise, don't raise the hips too high. Keep your abdominals engaged, which will ensure you don't arch your back excessively.

HAMSTRING WALKOUT

This movement targets the posterior chain muscles at the back of the body. The walkout motion particularly engages the hamstrings and the glutes.

KEY		
● Primary target muscle	● Secondary target muscle	

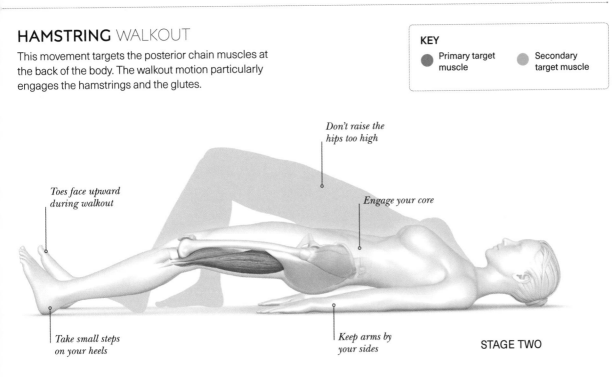

Don't raise the hips too high

Toes face upward during walkout

Engage your core

Take small steps on your heels

Keep arms by your sides

STAGE TWO

PREPARATORY STAGE
Begin in the starting position for a basic glute bridge (see pp.120–121). Tighten your abs by pushing your lower back into the floor and squeezing the glute muscles before you push up.

STAGE ONE
Exhale and raise up into an isometric glute bridge, squeezing the glutes and fully engaging the core.

STAGE TWO
Lift the toes off the floor, then take small, deliberate steps as you walk on the heels of your feet back and forth. Maintain the bridge as you walk. Take 2–4 steps out and 2–4 steps back in.

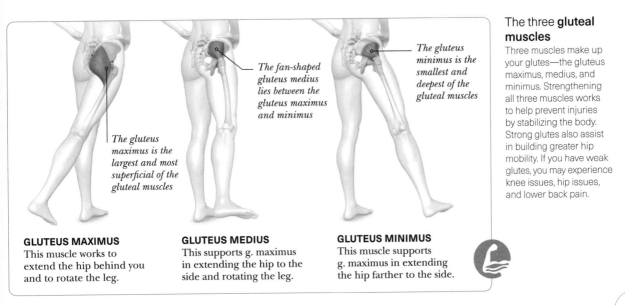

The fan-shaped gluteus medius lies between the gluteus maximus and minimus

The gluteus minimus is the smallest and deepest of the gluteal muscles

The gluteus maximus is the largest and most superficial of the gluteal muscles

GLUTEUS MAXIMUS
This muscle works to extend the hip behind you and to rotate the leg.

GLUTEUS MEDIUS
This supports g. maximus in extending the hip to the side and rotating the leg.

GLUTEUS MINIMUS
This muscle supports g. maximus in extending the hip farther to the side.

The three **gluteal muscles**

Three muscles make up your glutes—the gluteus maximus, medius, and minimus. Strengthening all three muscles works to help prevent injuries by stabilizing the body. Strong glutes also assist in building greater hip mobility. If you have weak glutes, you may experience knee issues, hip issues, and lower back pain.

PLYOMETRIC EXERCISES

Plyometric exercises are explosive, powerful, quick movements that exert maximum effort in a short amount of time. These exercises are aimed at increasing the heart rate, as well as improving stability, muscular strength, agility, cardiovascular strength and endurance, flexibility, and athletic performance. It's very important to warm up prior to performing plyometric moves—without an adequate warm-up, you could cause injury. Plan on doing these in the middle or at the end of a HIIT workout.

SKATER

This advanced cardiovascular exercise increases your fitness and strength. It also works your legs—mostly the quadriceps and the glutes, as well as the hamstrings and calves. Stability and balance are improved by always keeping the core engaged. The outer glute is a main focus here.

THE BIG PICTURE

Make sure that the area where you plan to do the skater exercise is free of obstacles and that you are jumping as far as you can to each side; small jumps are ineffective. The motion of the arms helps propel you back and forth. The back leg always goes behind the front leg; you are not doing a side-to-side shuffle. To start, perform for 30–60 seconds. For a more advanced version, don't tap your toe on the floor as you bring it back.

KEY

•-- *Joints*

○— *Muscles*

⬤ Shortening with tension

⬤ Lengthening with tension

⬤ Lengthening without tension

⬤ Held muscles without motion

ANTERIOR VIEW

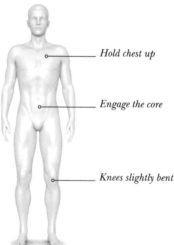

Hold chest up

Engage the core

Knees slightly bent

PREPARATORY STAGE

Start standing with feet shoulder-width apart, knees slightly bent, chest up, looking straight ahead, with head, neck, and spine in alignment. Arms are relaxed by your side. If you are starting on the left side of your mat, that means you'll be jumping to the right side.

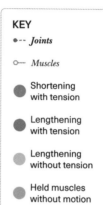

Rectus femoris
Gracilis
Vastus medialis
Sartorius
Gastrocnemius
Tibialis anterior

Lower **body**

Skaters target the muscles in the **glutes**, **quadriceps**, **hamstrings**, and **calves**. Your quadriceps are important in this movement because they extend each leg at the knee and flex your leg at the hip. Your glutes promote movement at your hip, helping extend, rotate, abduct, and adduct your leg.

STAGE ONE

Standing on your right foot, with a slight bend in the right knee, hop as far as you can to the left. Tighten your core to keep your balance as you use the momentum of your arms to drive off the standing leg to hop to the side.

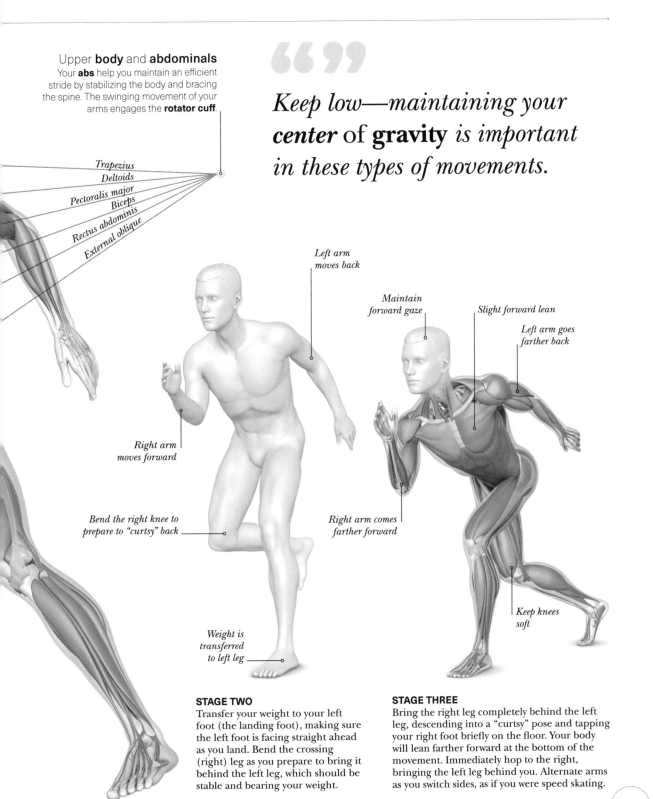

Upper **body** and **abdominals**

Your **abs** help you maintain an efficient stride by stabilizing the body and bracing the spine. The swinging movement of your arms engages the **rotator cuff**.

Trapezius
Deltoids
Pectoralis major
Biceps
Rectus abdominis
External oblique

> *Keep low—maintaining your center of gravity is important in these types of movements.*

Left arm moves back

Maintain forward gaze

Slight forward lean

Left arm goes farther back

Right arm moves forward

Bend the right knee to prepare to "curtsy" back

Right arm comes farther forward

Keep knees soft

Weight is transferred to left leg

STAGE TWO

Transfer your weight to your left foot (the landing foot), making sure the left foot is facing straight ahead as you land. Bend the crossing (right) leg as you prepare to bring it behind the left leg, which should be stable and bearing your weight.

STAGE THREE

Bring the right leg completely behind the left leg, descending into a "curtsy" pose and tapping your right foot briefly on the floor. Your body will lean farther forward at the bottom of the movement. Immediately hop to the right, bringing the left leg behind you. Alternate arms as you switch sides, as if you were speed skating.

127

HIGH KNEE

The high knee is both a strength and cardiovascular movement and is a great exercise to perform as a warm-up or as part a HIIT workout. As well as building cardiovascular endurance, this type of metabolic workout helps you burn fat.

THE **BIG PICTURE**

Start slowly at first (warm-up speed) and build up cardiovascular endurance by taking the speed up incrementally. Keep the back straight, with the head, spine, and neck all in alignment. As the knee comes up, the back is still straight and the chest is open. Use the arms to help get the knee up as high as you can. Move the arms, bent at a 90-degree angle, as if you were running, alternating opposite arm and knee. Alternate the knees for 30–60 seconds, speeding up as you become stronger and fitter.

Lower **body**
The **hip flexors, calves, glutes, quadriceps,** and **hamstrings** are recruited in the high knee. You isometrically tone your calves, quadriceps, hamstrings, and glutes on the standing leg and the calf contracts on the raised leg when you push off the floor with your foot.

⚠ Caution
As with all plyometric exercises, make sure you are physically fit enough for this type of movement.

KEY
- ●-- *Joints*
- ○— *Muscles*
- ● Shortening with tension
- ● Lengthening with tension
- ● Lengthening without tension
- ● Held muscles without motion

LATERAL-ANTERIOR VIEW

Upper **body** and **abdominals**
Because your arms are pumping, you may feel tension in your **shoulders** and **arms**. The **abdominals** work to create balance, as you are standing on one leg at a time. They also help stabilize the body and hold the spine in a neutral position.

Sternocleidomastoid
Deltoids
Pectoralis major
Biceps
Rectus abdominis
External oblique

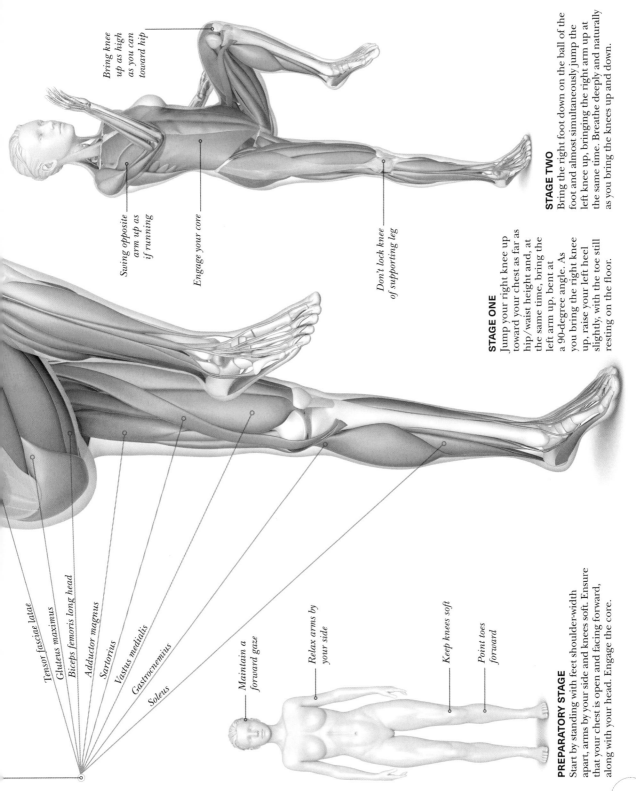

Bring knee up as high as you can toward hip

Swing opposite arm up as if running

Engage your core

Don't lock knee of supporting leg

Tensor fasciae latae

Gluteus maximus

Biceps femoris long head

Adductor magnus

Sartorius

Vastus medialis

Gastrocnemius

Soleus

Maintain a forward gaze

Relax arms by your side

Keep knees soft

Point toes forward

STAGE ONE

Jump your right knee up toward your chest as far as hip/waist height and, at the same time, bring the left arm up, bent at a 90-degree angle. As you bring the right knee up, raise your left heel slightly, with the toe still resting on the floor.

STAGE TWO

Bring the right foot down on the ball of the foot and almost simultaneously jump the left knee up, bringing the right arm up at the same time. Breathe deeply and naturally as you bring the knees up and down.

PREPARATORY STAGE

Start by standing with feet shoulder-width apart, arms by your side and knees soft. Ensure that your chest is open and facing forward, along with your head. Engage the core.

129

» VARIATIONS

These high knee variations are cardio-intensive exercises aimed at getting the heart rate up. They engage multiple muscle groups, primarily the core, hip flexors, calves, quadriceps, and hamstrings. All of these variations are perfect for warming up the body before a workout. Adding the jump rope engages the biceps, forearms, and deltoids.

JUMP ROPE HIGH KNEE

This is an advanced cardiovascular endurance exercise, focusing on strengthening the legs, especially the calves. You will need a jump rope to perform this exercise. Start the high knee exercise slowly and get into a rhythm.

*Don't perform **jump rope exercises** on hard surfaces—this can harm the knees and shins and lead to pain or injuries.*

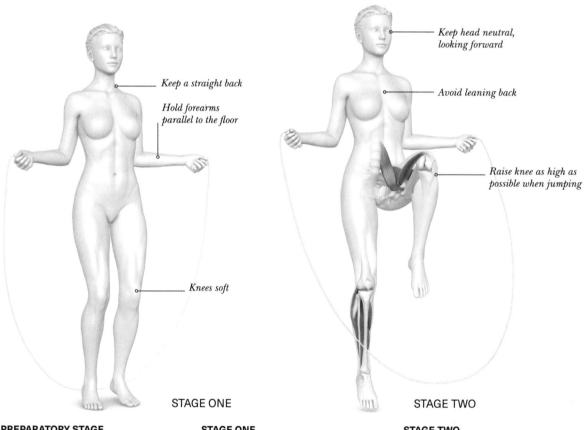

Keep a straight back

Hold forearms parallel to the floor

Knees soft

STAGE ONE

Keep head neutral, looking forward

Avoid leaning back

Raise knee as high as possible when jumping

STAGE TWO

PREPARATORY STAGE
Stand on a soft surface, such as a gym mat, feet slightly apart. Hold a rope handle in each hand and place your feet in front of the jump rope.

STAGE ONE
Tuck your elbows into your sides and bring your hands up. Swing the rope up and over your head, then jump up off the floor one foot at a time.

STAGE TWO
Raise each knee as high as possible when jumping. Repeat the exercise for 30–60 seconds, alternating knees with each turn of the rope.

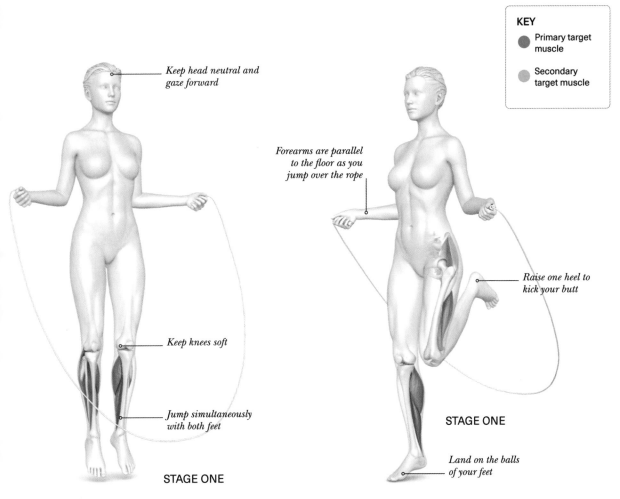

KEY
● Primary target muscle
● Secondary target muscle

Keep head neutral and gaze forward

Keep knees soft

Jump simultaneously with both feet

STAGE ONE

Forearms are parallel to the floor as you jump over the rope

Raise one heel to kick your butt

STAGE ONE

Land on the balls of your feet

JUMP ROPE FEET TOGETHER

This is similar to the jump rope high knee, but this time you are jumping over the rope with both feet at the same time.

PREPARATORY STAGE
Stand on a soft surface, such as a gym mat, feet slightly apart. Hold a rope handle in each hand and place your feet in front of the jump rope so that the rope is resting on the floor behind your heels.

STAGE ONE
Tuck your elbows into your sides and bring your hands up so that your forearms are parallel to the floor. Swing the rope up and over your head, then jump up off the ground with both your feet together just before the rope reaches your feet.

STAGE TWO
Repeat the exercise for 30–60 seconds.

JUMP ROPE BUTT KICK

Butt kicks increase the speed of hamstring contractions, which can help you run faster. Adding the jump rope increases the calf muscle engagement.

PREPARATORY STAGE
Stand on a soft surface, feet slightly apart. Hold a rope handle in each hand and place feet in front of the jump rope so that it is on the floor behind your heels.

STAGE ONE
Tuck your elbows into your sides and bring your hands up so that your forearms are parallel to the floor. Swing the rope up and over your head, then jump over the rope one foot at a time, raising the heel to kick your butt as the rope comes to meet them.

STAGE TWO
Try to touch the butt each time. Repeat for 30–60 seconds.

SQUAT JUMP

The squat (or plyometric) jump improves agility, balance, and power, as well as helping improve an athlete's vertical jump. It strengthens the glutes abdominals, hamstrings, and lower back.

THE **BIG PICTURE**

Due to the explosive nature of this exercise, it's crucial to have your body warmed up before performing it, so don't attempt it at the beginning of a workout. Engage your core to avoid straining your lower back, and distribute your weight evenly when you land. Start with 1 set of 5–10 reps. Slowly work yourself up to 3 sets.

KEY

- ●-- *Joints*
- ○- *Muscles*
- ● Shortening with tension
- ● Lengthening with tension
- ● Lengthening without tension
- ● Held muscles without motion

Abdominals **and** upper **body**
The **erector spinae** help rotate and extend the spine and neck. The **rectus abdominis**, **obliques**, and **transverse abdominis** stabilize the torso during the jump, helping maintain a straight spine. Using the swinging motion in the jump adds tension to the arms and shoulders.

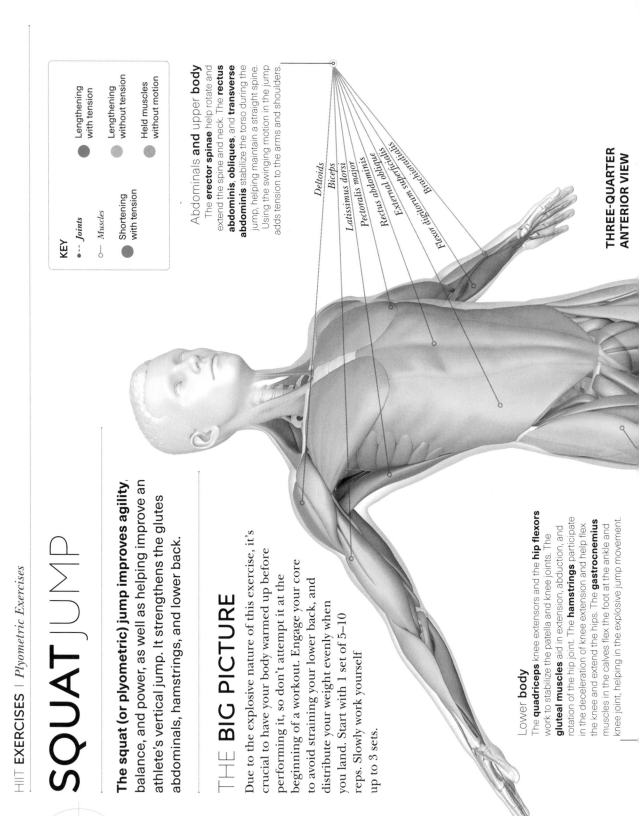

Deltoids
Biceps
Latissimus dorsi
Pectoralis major
Rectus abdominis
External oblique
Flexor digitorum superficialis
Brachioradialis

THREE-QUARTER ANTERIOR VIEW

Lower **body**
The **quadriceps** knee extensors and the **hip flexors** work to stabilize the patella and knee joints. The **gluteal muscles** aid in extension, abduction, and rotation of the hip joint. The **hamstrings** participate in the deceleration of knee extension and help flex the knee and extend the hips. The **gastrocnemius** muscles in the calves flex the foot at the ankle and knee joint, helping in the explosive jump movement.

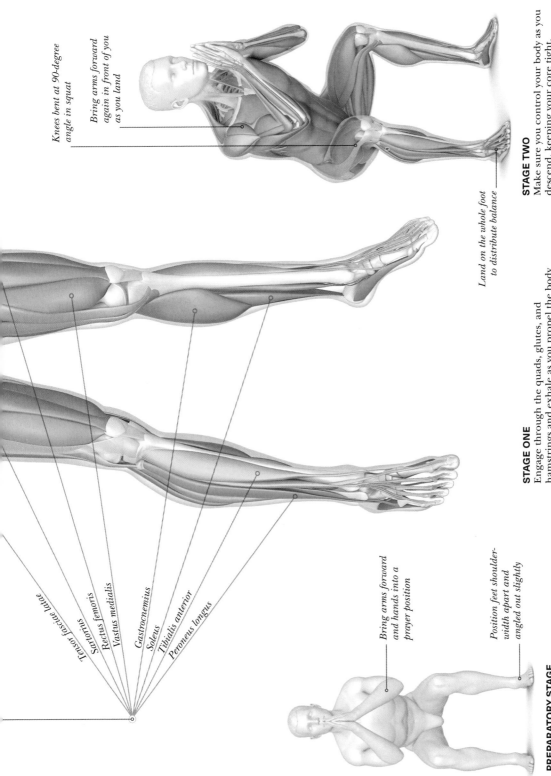

Knees bent at 90-degree angle in squat

Bring arms forward again in front of you as you land

Land on the whole foot to distribute balance

STAGE TWO
Make sure you control your body as you descend, keeping your core tight. Control your landing by going through your foot (toes, ball, arches, heel) and descend into the squat again for another explosive jump. Upon landing, immediately repeat the jump.

STAGE ONE
Engage through the quads, glutes, and hamstrings and exhale as you propel the body up and off the floor into an explosive jump, extending through the legs. With the legs fully extended, the feet will be airborne off the floor. Bring your arms out to the side in front as you jump to help propel you upward.

Tensor fasciae latae
Sartorius
Rectus femoris
Vastus medialis
Gastrocnemius
Soleus
Tibialis anterior
Peroneus longus

Bring arms forward and hands into a prayer position

Position feet shoulder-width apart and angled out slightly

PREPARATORY STAGE
Stand with feet shoulder-width apart and knees slightly bent. Engaging your core, bend your knees and descend into a full squat position, knees bent and thighs parallel to the floor.

» VARIATIONS

These squat jump variations are plyometric (explosive power) movements that increase cardiovascular and muscular endurance. Your abdominals, glutes, hamstrings, and lower back muscles are all engaged here. You shouldn't do these every day—allow your body 48–72 hours to recover.

66 99

*In the frog and sumo jumps, don't let your **knees** cave inward when jumping from the wider starting point.*

KEY

● Primary target muscle

● Secondary target muscle

FROG JUMP

The frog jump strengthens the quadriceps, glutes, calves, hamstrings, inner thighs, and hip flexors. It's a plyometric exercise geared toward improving muscle mass, speed, and agility.

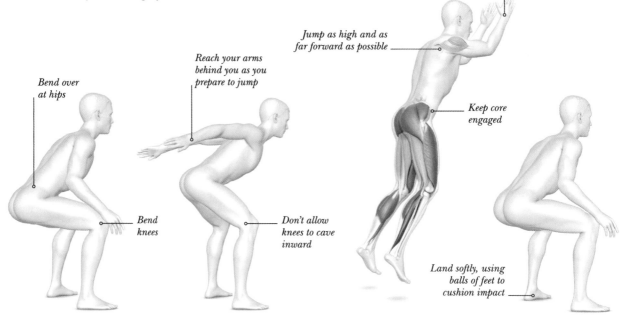

Bend over at hips

Bend knees

Reach your arms behind you as you prepare to jump

Don't allow knees to cave inward

Jump as high and as far forward as possible

Move arms forward as you jump

Keep core engaged

Land softly, using balls of feet to cushion impact

PREPARATORY STAGE
Stand on one side of your mat with legs wide and toes pointed forward, ensuring there is space in front of you. Descend into a wide squat.

STAGE ONE
Tightening your core, prepare to jump up and forward. Make sure you're low at the start of the jump, in a "froglike" position.

STAGE TWO
Jump up and forward, propelling yourself as far as possible, moving your arms forward to help your momentum.

STAGE THREE
Land on your toes and then the balls of your feet in a low squat resembling that of a frog. Turn around and jump back in the same way.

IN-AND-OUT SQUAT **JUMP**

This squat jump variation really focuses on speed, agility, and power. It works the abs, glutes, hamstrings, and lower back muscles. Additionally, it targets the calves, adductors, and abductors.

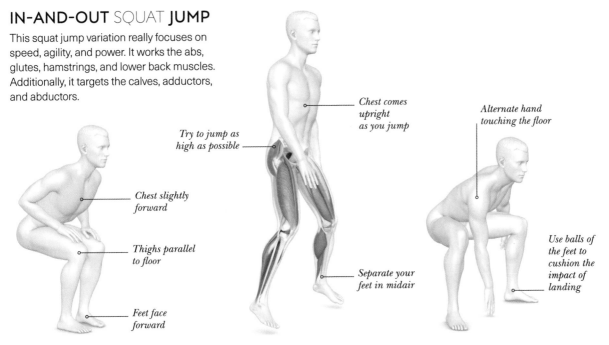

Chest comes upright as you jump

Try to jump as high as possible

Chest slightly forward

Thighs parallel to floor

Feet face forward

Separate your feet in midair

Alternate hand touching the floor

Use balls of the feet to cushion the impact of landing

PREPARATORY STAGE
Stand with feet together, knees slightly bent, with your hands in front of you resting on your thighs. Inhale as you descend into a chair squat (see p.98).

STAGE ONE
Explosively jump, and as you are coming down, separate your feet in midair. Land with feet wide apart, lowering into a squat and touching one hand on the floor.

STAGE TWO
Land softly, passing the jump through your toes, the balls of the feet, then to the heels. Jump up out of the squat and land back in the starting position.

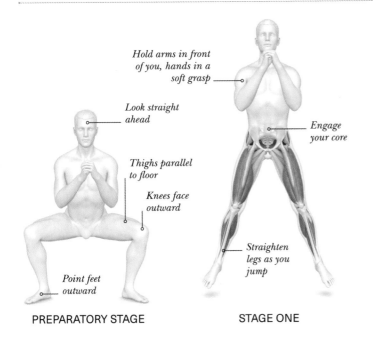

Hold arms in front of you, hands in a soft grasp

Look straight ahead

Thighs parallel to floor

Knees face outward

Engage your core

Point feet outward

Straighten legs as you jump

PREPARATORY STAGE

STAGE ONE

SUMO SQUAT **JUMP**

This plyometric jump sculpts the legs and the posterior. It engages the glutes and inner and outer thighs, as well as the abdominals, quadriceps, hamstrings, and lower back.

PREPARATORY STAGE
Start in a wide stance, legs wide and feet pointed outward. Slowly bend at the hips and knees into a deep sumo squat (see p.99).

STAGE ONE
Exhale, and using the glutes, hamstrings, quadriceps, and core, propel your body up into the air. Straighten both your legs and your hips in midair before you come back to the floor.

STAGE TWO
Land back into the sumo squat position. When landing, ensure that you maintain soft knees to prevent injury, and use the balls of your feet to cushion the impact. Repeat for 30–60 seconds.

TUCK JUMP

In the tuck jump, you use your body weight and power to contract several muscles at once to jump up in the air. It requires strength and cardiovascular endurance and strengthens your quadriceps, glutes, hamstrings, calves, hip flexors, abdominals, and obliques.

THE BIG PICTURE

Make sure you don't perform this at the very beginning of a HIIT workout—you need to be warmed up before this and any plyometric exercise, or you can strain the knees and joints. It's important to know how to land when doing these plyometric jumps: make sure you are landing softly through the feet, knees, and hips. Use the full range of body motion when you jump. Complete 1 set of 4–8 reps if beginning. This is an advanced fitness move, so don't perform more than twice a week to avoid too much joint impact.

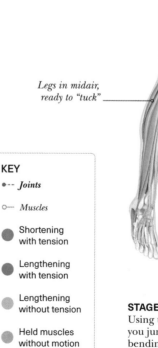

Swing arms up to help momentum of jump

Keep lower abs engaged to drive your knees up

Legs in midair, ready to "tuck"

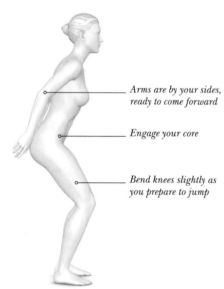

Arms are by your sides, ready to come forward

Engage your core

Bend knees slightly as you prepare to jump

PREPARATORY STAGE
Start in the standing position with feet hip-width apart, arms by your sides, knees soft, and core engaged. Bend the knees slightly in preparation for driving yourself off the floor into the jump.

KEY

●-- *Joints*

○-- *Muscles*

● Shortening with tension

● Lengthening with tension

● Lengthening without tension

● Held muscles without motion

STAGE ONE
Using the leg muscles, exhale as you jump straight up into the air, bending the arms and bringing them forward and up.

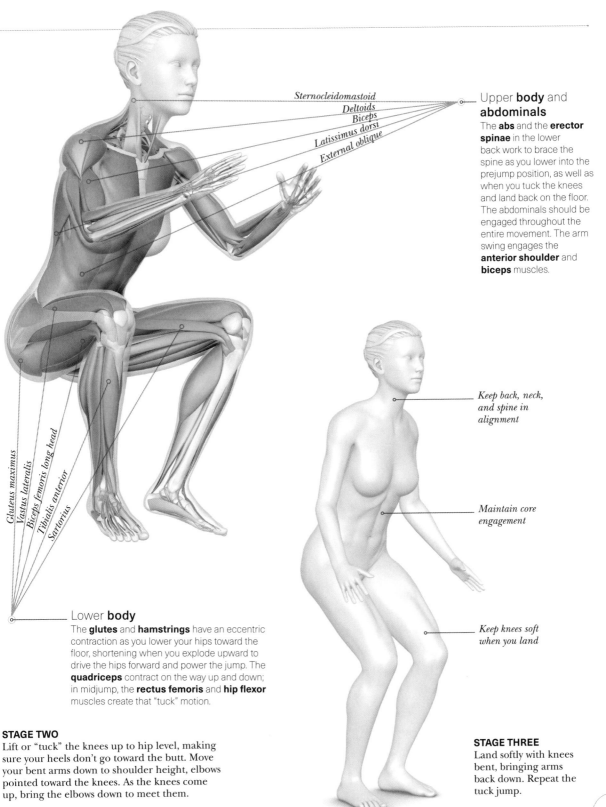

Sternocleidomastoid
Deltoids
Biceps
Latissimus dorsi
External oblique

Upper **body** and abdominals

The **abs** and the **erector spinae** in the lower back work to brace the spine as you lower into the prejump position, as well as when you tuck the knees and land back on the floor. The abdominals should be engaged throughout the entire movement. The arm swing engages the **anterior shoulder** and **biceps** muscles.

Gluteus maximus
Vastus lateralis
Biceps femoris long head
Tibialis anterior
Sartorius

Keep back, neck, and spine in alignment

Maintain core engagement

Keep knees soft when you land

Lower **body**

The **glutes** and **hamstrings** have an eccentric contraction as you lower your hips toward the floor, shortening when you explode upward to drive the hips forward and power the jump. The **quadriceps** contract on the way up and down; in midjump, the **rectus femoris** and **hip flexor** muscles create that "tuck" motion.

STAGE TWO
Lift or "tuck" the knees up to hip level, making sure your heels don't go toward the butt. Move your bent arms down to shoulder height, elbows pointed toward the knees. As the knees come up, bring the elbows down to meet them.

STAGE THREE
Land softly with knees bent, bringing arms back down. Repeat the tuck jump.

137

BOX JUMP

The box jump is a plyometric workout that targets all of the muscle groups of your lower body, including your glutes, hamstrings, quadriceps, and calves. Because you also engage the core and use the arms in a swinging motion, it's really a full-body workout.

THE **BIG PICTURE**

The key to mastering the box jump is to make sure you start with a box appropriate for your current fitness level. If you are a beginner, you'll want to start with a low box 12in (30cm) high so you can get accustomed to the movement. You can progress to a higher box as you gain confidence. Plan to do 10–12 reps for 3 sets if just beginning.

STAGE TWO
Jump straight up into the air, exploding through the balls of your feet. Swing your arms up and forward you fully extend your knees and hips to get much height as you c At the top of the jum bend your knees and hips to draw them upward as you land o top of the box.

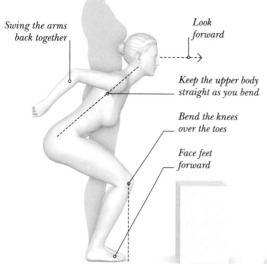

Swing the arms back together

Look forward

Keep the upper body straight as you bend

Bend the knees over the toes

Face feet forward

Biceps
Triceps
Deltoids
Pectoralis major
Latissimus dorsi
Serratus anterior
External oblique
Rectus abdominis

Tensor fasciae latae
Rectus femoris
Hip
Adductor magnus
Biceps femoris (l.h.)
Vastus medialis
Knee
Gastrocnemius
Tibialis anterior
Peroneus longus
Ankle
Abductor digiti minimi
Extensor digitorum longus

Upper **body** and **arms**
The movement of the **arms** assists in bringing the body off the floor, creating the momentum needed to propel the body upward. The **rectus abdominis** and **obliques** come into play as the body elongates when you jump.

Legs
The **quadriceps** muscles work together to extend your knees. The calf muscles, the **gastrocnemius** and **soleus**, create the spring motion in the jump. **Hamstring** muscles work together to flex your knee and extend your hip. The **glutes** help in hip extension.

STAGE ONE
Start by standing facing the box. Your feet should be about hip-distance apart, your knees and hips slightly bent in an athletic stance. Bend your knees and press your hips back as you swing your arms behind you.

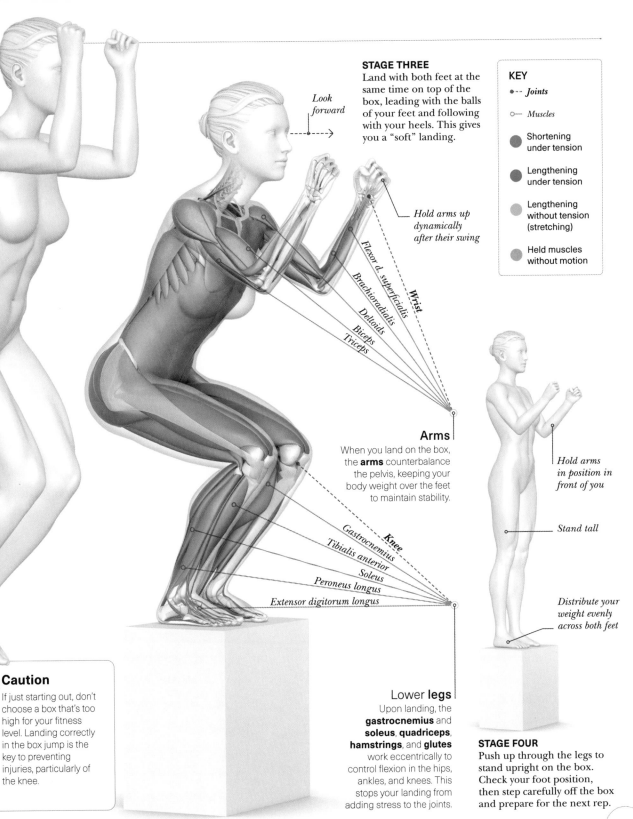

STAGE THREE
Land with both feet at the same time on top of the box, leading with the balls of your feet and following with your heels. This gives you a "soft" landing.

Look forward

Hold arms up dynamically after their swing

KEY

- •--- *Joints*
- ○— *Muscles*
- ● Shortening under tension
- ● Lengthening under tension
- ● Lengthening without tension (stretching)
- ● Held muscles without motion

Flexor d. superficialis

Brachioradialis

Deltoids

Biceps

Triceps

Wrist

Arms
When you land on the box, the **arms** counterbalance the pelvis, keeping your body weight over the feet to maintain stability.

Knee

Gastrocnemius

Tibialis anterior

Soleus

Peroneus longus

Extensor digitorum longus

Hold arms in position in front of you

Stand tall

Distribute your weight evenly across both feet

Caution
If just starting out, don't choose a box that's too high for your fitness level. Landing correctly in the box jump is the key to preventing injuries, particularly of the knee.

Lower **legs**
Upon landing, the **gastrocnemius** and **soleus**, **quadriceps**, **hamstrings**, and **glutes** work eccentrically to control flexion in the hips, ankles, and knees. This stops your landing from adding stress to the joints.

STAGE FOUR
Push up through the legs to stand upright on the box. Check your foot position, then step carefully off the box and prepare for the next rep.

139

SINGLE-LEG FORWARD JUMP

Upper **body** and **abdominals**
The **transverse abdominis**, **rectus abdominis**, and **internal** and **external obliques** help brace the spine, holding it in the neutral position, as well as stabilizing the torso. The arm swing engages the **anterior shoulders**, **rotator cuff**, and **biceps**.

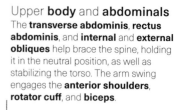

Trapezius
Deltoids
Latissimus dorsi
Rectus abdominis
External oblique

This powerful plyometric move strengthens the calves, glutes, hip flexors, hamstrings, and quadriceps. It improves agility, speed, balance, and overall athletic performance.

THE **BIG PICTURE**

Make sure the area all around you is clear before you begin this exercise. As with all jumps, how you land is important. Avoid any twisting or sideways motion at the knee and the ankle. Complete 3–10 reps on one leg, then switch sides. Plyometric drills like this should only be performed twice a week to give muscles a chance to recover.

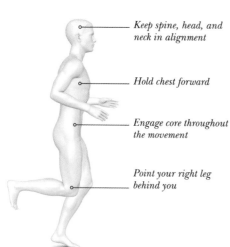

Keep spine, head, and neck in alignment

Hold chest forward

Engage core throughout the movement

Point your right leg behind you

Arms move forward and back as you jump

Right leg comes forward

Hop on the ball of your foot

PREPARATORY STAGE
Start by standing up straight with the legs shoulder-width apart, holding the back straight and pushing out the chest. Lift your right leg off the floor, pointing it behind you. Bend your left knee slightly and push against the floor to hop up and forward.

STAGE ONE
Bring your hopping (left) leg forward slightly at the knee to help with the momentum of the leap. Use the right leg to help you move the body forward. As you propel forward, the arms move back as well.

HEAD

●--- *Joints*

○--- *Muscles*

● Shortening with tension

● Lengthening with tension

● Lengthening without tension

● Held muscles without motion

Lower **body**

The muscles in the **calves**, **hamstrings**, **quadriceps**, and **hip flexors** are engaged here. The quads contract eccentrically on the way down from the hop and then concentrically to extend the knees as you jump. While hopping, the **rectus femoris** and the **hip flexor** muscles help propel the body forward.

Gluteus medius
Gluteus maximus
Bicep femoris long head
Adductor magnus
Rectus femoris
Vastus medialis
Gastrocnemius
Soleus
Abductor hallucis

Right arm comes back to help with leap momentum

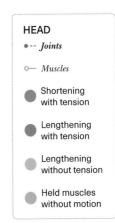

Bring right leg forward to complete another jump

Land on the balls of your feet

THREE-QUARTER LATERAL VIEW

STAGE TWO

Kick your right leg back as you leap forward, moving your bent arms back and forward as you hop to help with the movement.

STAGE THREE

Land softly on the toes and roll to the heels, which helps absorb the force of impact. Bring your right leg back to the front and hop forward again on the left leg, kicking your right leg out behind you. Repeat the movement, then switch legs.

FOOTBALL
UP AND DOWN

This cardiovascular and strength-building exercise is a combination of running in place and a burpee. It strengthens the abs, triceps, upper back, chest, shoulders, calves, and quadriceps. It also improves coordination and agility.

THE BIG PICTURE

Check that your surface is level before starting this exercise. It begins by keeping low with knees bent, but don't let the knees go past the toes as you jump down to the floor into the push-up. If you are a beginner, start with running in one spot 8 times before dropping down to the floor; repeat for 30 seconds.

KEY

●-- *Joints*

○-- *Muscles*

● Shortening with tension

● Lengthening with tension

● Lengthening without tension

● Held muscles without motion

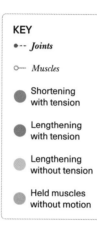

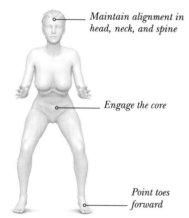

Maintain alignment in head, neck, and spine

Engage the core

Point toes forward

PREPARATORY STAGE
Stand with feet shoulder-width apart, crouched down slightly in a half-squat position. Bend your elbows to 90 degrees and keep them at this angle for stage 1.

STAGE ONE
Begin to run on the spot, quickly hopping from one foot to the other, remaining on the balls of your feet. Keep your arms static as you do this fast, explosive movement. Complete 8 quick, low hops without stopping.

ANTERIOR-LATERAL VIEW

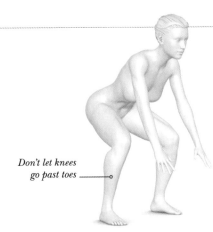

Don't let knees go past toes

STAGE TWO
After your 8th hop, crouch down and prepare to put your hands on the floor underneath your shoulders.

Jump feet out behind you to land on your toes

Transfer weight to your hands

STAGE THREE
Take your weight into your hands and shoulders and jump both of your feet out behind you to straighten your legs, keeping the core engaged.

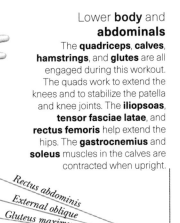

Lower **body** and **abdominals**
The **quadriceps**, **calves**, **hamstrings**, and **glutes** are all engaged during this workout. The quads work to extend the knees and to stabilize the patella and knee joints. The **iliopsoas**, **tensor fasciae latae**, and **rectus femoris** help extend the hips. The **gastrocnemius** and **soleus** muscles in the calves are contracted when upright.

Rectus abdominis
External oblique
Gluteus maximus
Rectus femoris
Gastrocnemius
Peroneus longus
Tibialis anterior

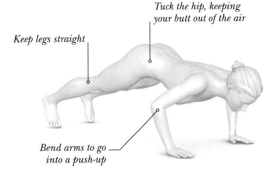

Tuck the hip, keeping your butt out of the air

Keep legs straight

Bend arms to go into a push-up

STAGE FOUR
Get into a high plank position, weight on your hands and toes, in preparation for doing a push-up.

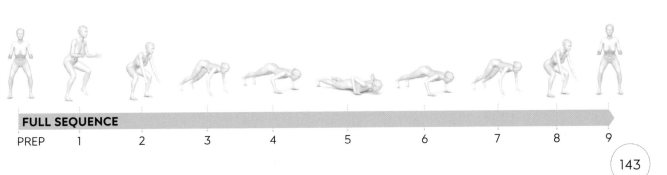

FULL SEQUENCE

PREP   1   2   3   4   5   6   7   8   9

143

» **FOOTBALL** UP AND DOWN (CONTINUED)

Flex your toes on the floor

Lift hands off the floor briefly

STAGE FIVE
Bend your elbows and lower your upper body to the floor into a push-up, chest in between your hands. Touch your chest to the floor and lift your palms up briefly. Rest the toes on the floor and keep your feet hip-width apart.

! Caution
This is a very quick-moving cardiovascular exercise. Make sure that you are physically ready and that you keep the core engaged to protect the back.

Upper **body** and **abdominals**
The **triceps**, **upper back**, **chest**, and **shoulders** are engaged in the fifth stage of the movement. The **erector spinae** helps extend the thoracic and lumbar spine. When you jump back, the **rectus abdominis** and **obliques** contract to brace the spine. The **pectoralis major**, **triceps brachii**, and **anterior deltoids** are activated, as well as the middle and upper **trapezius** muscles.

Latissimus dorsi
External oblique
Triceps
Deltoids
Trapezius

Abductor digiti minimi
Peroneus longus
Gastrocnemius
Vastus lateralis
Biceps femoris long head
Gluteus maximus
Gluteus medius

LATERAL VIEW

STAGE SIX
Press your palms back into the floor to push your upper body back up away from the floor, completing the press-up.

Lower **body**
In the burpee, the **quadriceps**, **glutes**, **hamstrings**, and **calves** are held isometrically to stabilize the body. The **gluteus maximus** and **hamstrings** work to flex the hip when you jump back. Hamstrings also aid in flexing the knees.

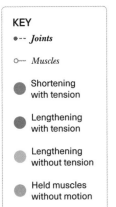

KEY

●-- *Joints*

○— *Muscles*

● Shortening with tension

● Lengthening with tension

● Lengthening without tension

● Held muscles without motion

"
*Maintaining a **low center of** gravity will help you move quickly and retain balance by keeping your **weight evenly** distributed.*

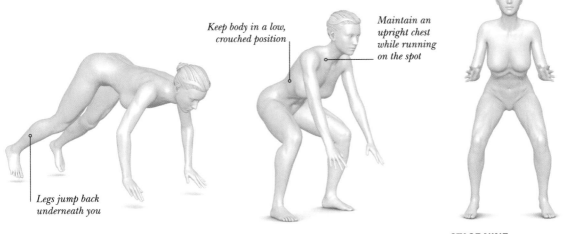

Keep body in a low, crouched position

Maintain an upright chest while running on the spot

Legs jump back underneath you

STAGE SEVEN
Jump your legs underneath you, preparing to come back to standing in the half-squat position, knees bent but soft.

STAGE EIGHT
Land with your feet shoulder-width apart. Transfer your body weight back onto your legs and bend your elbows into the preparatory stage position.

STAGE NINE
Run on the spot rapidly for 8 hops before jumping down again and repeating the sequence.

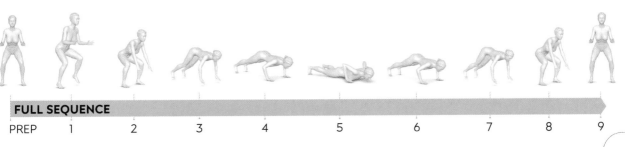

FULL SEQUENCE

PREP   1   2   3   4   5   6   7   8   9

145

BURPEE

This full-body workout strengthens both the lower and upper body. It increases agility, power, and endurance, targeting the legs, hips, glutes, abdomen, chest, shoulders, and arms. It's a high-intensity exercise aimed at helping get the heart rate up, which will also speed up your metabolism.

THE **BIG PICTURE**

The burpee is a very demanding exercise, combining explosive jumping with strength-building push-ups. To challenge yourself further, you can do a tuck jump instead of a jump with straight legs, bringing the knees up toward the chest. With a straight spine and neck and head in alignment, power into the tuck jump from the legs at stage one. Tighten the core to bring the knees in. Start with 5 reps, and build up to 10 as you become accustomed to the exercise.

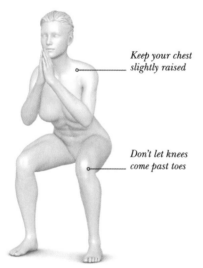

Keep your chest slightly raised

Don't let knees come past toes

PREPARATORY STAGE
Start in a squat: position your feet shoulder-width apart, spine and neck in alignment, and knees bent. Make sure the knees don't come past the toes and the chest remains no lower than 45-degree angle.

Flexor digitorum superficialis
Biceps
Pectoralis major
Deltoids
Rectus abdominis
External oblique

Upper **body**
The burpee engages a lot of the upper body because there is a push-up involving contraction of the **pectoralis major**, **deltoids**, and **triceps**. The **abdominal muscles** are also engaged, in bracing the spine. The **erector spinae** muscles hold the body stable. In the jump, swinging the arms outward engages the **shoulders**.

STAGE ONE
Using the power from your legs, jump quickly into the air and land back where you started. Swing your arms out to the side as you jump, keeping your legs straight.

LATERAL VIEW

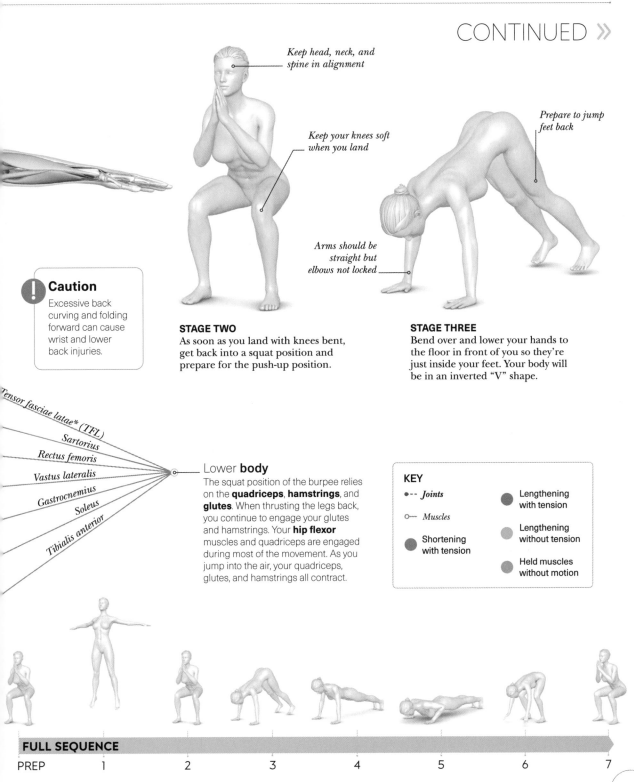

Keep head, neck, and spine in alignment

Keep your knees soft when you land

Arms should be straight but elbows not locked

Prepare to jump feet back

⚠ Caution

Excessive back curving and folding forward can cause wrist and lower back injuries.

STAGE TWO
As soon as you land with knees bent, get back into a squat position and prepare for the push-up position.

STAGE THREE
Bend over and lower your hands to the floor in front of you so they're just inside your feet. Your body will be in an inverted "V" shape.

Tensor fasciae latae* (TFL)
Sartorius
Rectus femoris
Vastus lateralis
Gastrocnemius
Soleus
Tibialis anterior

Lower **body**
The squat position of the burpee relies on the **quadriceps**, **hamstrings**, and **glutes**. When thrusting the legs back, you continue to engage your glutes and hamstrings. Your **hip flexor** muscles and quadriceps are engaged during most of the movement. As you jump into the air, your quadriceps, glutes, and hamstrings all contract.

KEY
- •-- *Joints*
- ○— *Muscles*
- ⬤ Shortening with tension
- ⬤ Lengthening with tension
- ⬤ Lengthening without tension
- ⬤ Held muscles without motion

FULL SEQUENCE

PREP   1   2   3   4   5   6   7

» BURPEE
(CONTINUED)

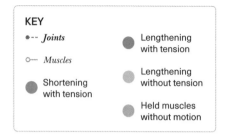

KEY

●-- *Joints*

○— *Muscles*

● Shortening with tension

● Lengthening with tension

● Lengthening without tension

● Held muscles without motion

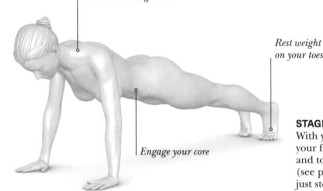

Keep your neck, spine, and head in alignment

Rest weight on your toes

Engage your core

STAGE FOUR

With your weight on your hands, jump your feet back so you're on your hands and toes and in a high plank position (see pp.34–35). To modify this, you can just step one foot back at a time.

Extensor digitorum

Pectoralis major

Deltoid

Biceps

Triceps

External oblique

Latissimus dorsi

Core **and** upper **body**

When performing the push-up, it's important to make sure the hip is tucked, the abdominals are engaged, and the elbows are slightly back when lowering so as to protect the deltoids.

STAGE FIVE

Keeping your body straight and core tight, do one push-up. Elbows come out toward the ceiling, chest to floor. Squeeze the thighs and don't let your back sag or stick your butt in the air when pushing up.

Land softly on
your feet

Arms come off the floor

STAGE SIX
Jump both your feet back to
their starting position, feet
planted firmly on the floor.

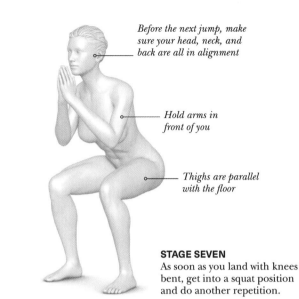

Before the next jump, make
sure your head, neck, and
back are all in alignment

Hold arms in
front of you

Thighs are parallel
with the floor

STAGE SEVEN
As soon as you land with knees
bent, get into a squat position
and do another repetition.

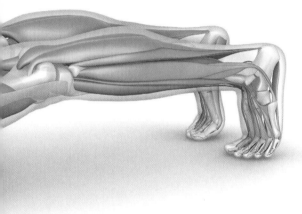

66 99

*The burpee speeds up
your metabolism, so you'll
continue to **burn** calories
throughout the day.*

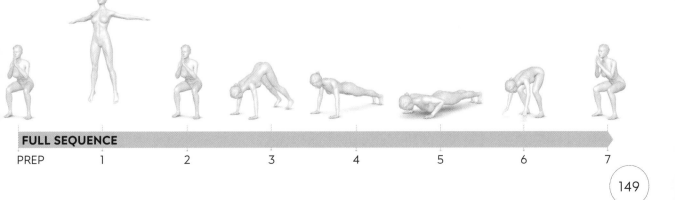

FULL SEQUENCE

PREP   1   2   3   4   5   6   7

BEAR CRAWL

The bear crawl is a mobility-focused core exercise. The total body movement in the bear crawl improves coordination and cardiovascular strength and endurance, as well as enhancing overall athletic performance. This exercise strengthens the shoulders, chest and back, glutes, quadriceps, hamstrings, and core.

Keep back strong

Feet hip-distance apart

Engage the core

THE **BIG PICTURE**

Keep the back completely flat as you propel your body forward. Before you start moving, brace your core so that the hips and shoulders are in one straight line. Maintain this solid core position as you move. Practice holding a tabletop position with the knees off the floor. The head should not sag forward or droop—that takes the head and neck out of alignment. Try to keep all of your movements underneath your torso as you move. If you notice your legs sneaking out to the side or hips swaying, you might be taking steps that are too big. If just beginning, perform the workout for 30 seconds. Slowly work up to 1–2 minutes for 3–5 reps forward and backward.

PREPARATORY STAGE

Get into a high plank (see pp.36–37) as if you were preparing to do a push-up. Hands should be beneath the shoulders, the back strong, and the core engaged, with feet hip-distance apart and heels off the floor.

Lower **body**

The bear crawl engages the muscles in the **quadriceps**, **glutes**, **hip flexors**, and **hamstrings**. The glutes help maintain stability in the hips as you move forward and backward, while the quadriceps are held isometrically throughout.

Gluteus maximus

Tensor fasciae latae

Biceps femoris long head

Rectus femoris

Vastus lateralis

Gastrocnemius

Soleus

Tibialis anterior

Peroneus longus

STAGE ONE

Transfer into a bear plank position (see pp.44–45). Begin to move forward by simultaneously moving the right hand and the left leg forward with a crawling motion. Immediately switch sides and move the left hand and right leg forward. Keep the body low as you crawl with small steps.

LATERAL VIEW

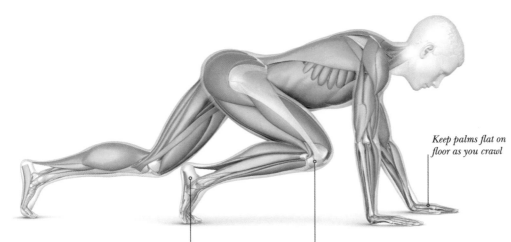

Keep palms flat on floor as you crawl

Keep heels off the floor and crawl on your toes

Knees at a 90-degree angle hovering off floor

STAGE TWO

Keeping your back straight and your knees bent, hovering about 2in (5cm) off the floor, repeat the crawling movement in reverse for the same amount of steps. Reset to the high plank position before you go into another cycle of reps.

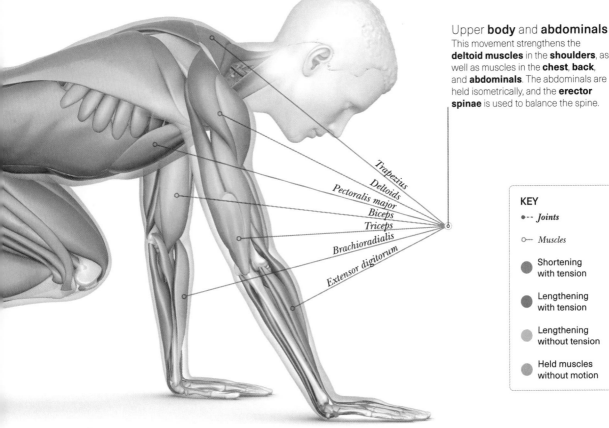

Upper **body** and **abdominals**

This movement strengthens the **deltoid muscles** in the **shoulders**, as well as muscles in the **chest**, **back**, and **abdominals**. The abdominals are held isometrically, and the **erector spinae** is used to balance the spine.

Trapezius
Deltoids
Pectoralis major
Biceps
Triceps
Brachioradialis
Extensor digitorum

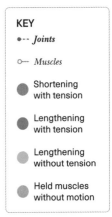

KEY

●·· *Joints*

○— *Muscles*

● Shortening with tension

● Lengthening with tension

● Lengthening without tension

● Held muscles without motion

151

TOTAL BODY
EXERCISES

Exercises in this section engage both the upper and lower body.
Each is an aerobic, resistance, and body weight compound exercise
aimed to cut time and burn more calories. Most of the exercises are
built by combining two main parts, making up a sequence that
will challenge the whole body. There are detailed instructions on
achieving proper form and minimizing risk.

JACK PRESS

The jack press is a cardiovascular endurance and strength-building exercise. Adding the press focuses on muscle endurance and power. This exercise strengthens the glutes, quads, hip flexors, and deltoids (shoulder muscles).

THE **BIG PICTURE**

The knees must stay soft and the heels off the floor, as you jump back and forth on the balls of your feet. Throughout this entire workout, it's important to keep the core engaged. Perform the workout for 30 seconds to start with. Slowly work up to 1–2 minutes for 3–5 reps, increasing the weights as you become accustomed to the exercise.

Upper body and **abdominals**
In the press, you use the **anterior** and **medial deltoid** to press straight up. The press movement also activates the **triceps** and **trapezius**. Lowering your arms back to your sides utilizes your **latissimus dorsi** in the back. The **rectus abdominis** prevents the spine from rounding, while the **obliques** work to keep you from leaning too far to the left or right.

Flexor digitorum superficialis

Brachioradialis

Biceps

Deltoids

Latissimus dorsi

Pectoralis major

Arms return to start of military press position

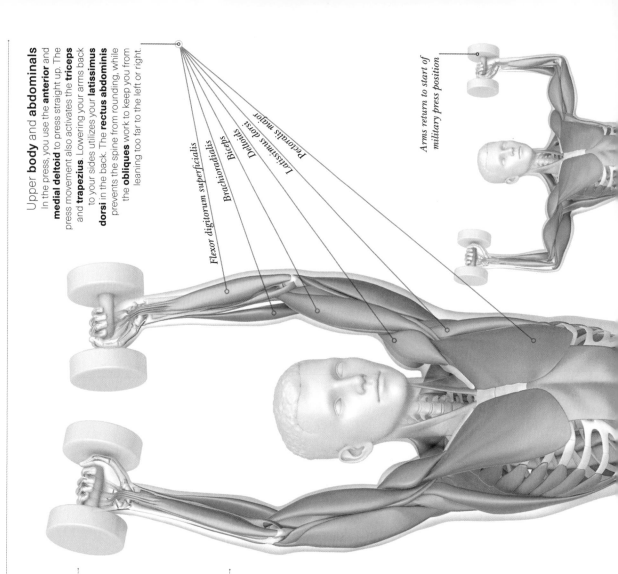

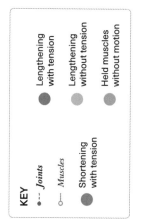

KEY

● - - Joints

○— Muscles

● Shortening with tension

● Lengthening with tension

● Lengthening without tension

● Held muscles without motion

Land on balls of the feet with your weight distributed evenly

STAGE TWO

Jump the feet back to their starting position while bringing the arms back down to their starting press position, making sure your elbows don't dip past the ribcage.

! Caution

Since you are jumping with weights, be careful not to use weights that are too heavy. Sometimes when doing the jack press, the arms give out and the back starts to cave in. Make sure that the spine, neck, and head are all in alignment and the core is engaged throughout the sequence.

ANTERIOR VIEW

Lower body

The **calf muscles** work to flex your ankles. The **quadriceps** extend your knees to propel you off the floor, and the **glutes** and **hamstrings** contract to extend your hips for the jump. Jumping out activates the **gluteus medius, gluteus minimus, tensor fasciae latae,** and **sartorius.** Jumping back engages the **adductors, pectineus,** and **gracilis** muscles.

Gracilis

Sartorius

Rectus femoris

Vastus medialis

Gastrocnemius

Tibialis anterior

Soleus

Hold weights horizontally, facing out

Bend elbows at 90 degrees to arm

Gentle bend in knees

PREPARATORY STAGE

Start by standing tall, feet shoulder-width apart, hands holding weights at your sides. The toes are pointed forward and the core is engaged. Lift the weights up into a military shoulder press position (see pp. 76–77), weights held horizontally; knuckles should face forward.

STAGE ONE

As you lift the weights above your head, jump your feet out to either side simultaneously in a jumping jack movement. The feet will jump out wider than shoulder-width apart.

PUSH-UP AND SQUAT

This combination full-body workout strengthens the chest muscles, shoulders, back of your arms, abdominals, and "wing" muscles directly under your armpit. It also strengthens all the largest muscles of the legs and butt, including the quadriceps, glutes, and hamstrings.

THE BIG PICTURE

For the push-up, keep the abs engaged during the whole movement—think belly button to spine. Moving from the push to the squat, it's important to jump the feet in landing on the whole of the foot, not just the toes, to distribute balance evenly. In the squat, make sure your knees don't go past your toes.

Lower **body** and **legs**
The **quadriceps**, **gluteus maximus** and **medius**, **hamstrings**, and **calf and shin muscles** are held isometrically to aid in stabilizing the body.

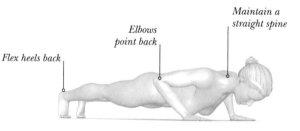

Gluteus maximus
Tensor fasciae latae
Rectus femoris
Vastus lateralis
Tibialis anterior
Peroneus longus

THREE-QUARTER ANTERIOR VIEW

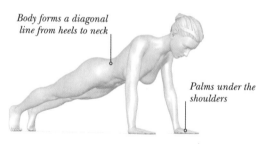

Body forms a diagonal line from heels to neck

Palms under the shoulders

PREPARATORY STAGE
Start in a high plank position (see pp.36–37) with your pelvis tucked in, neck neutral, and palms directly under your shoulders. Make sure your shoulders are rotated back and down and your core is engaged.

 Caution
Not engaging the core during the push-up can cause the spine to cave in, creating pressure on the lower back and joints. Incorrect form on the squat can lead to knee and lower back injuries. Avoid letting the knees bend inward, hunching the back, or lifting the heels off the floor.

Maintain a straight spine

Elbows point back

Flex heels back

STAGE ONE
Inhale, pulling the belly in and engaging the core. Keeping your back flat, exhale as you slowly lower your body by bending your elbows, until your chest grazes the floor.

KEY

- ●-- *Joints*
- ○— *Muscles*
- ● Shortening with tension
- ● Lengthening with tension
- ● Lengthening without tension
- ● Held muscles without motion

Upper **body** and **arms**

The upper body contracts as it lowers and raises the **pectoralis major and minor**, **deltoids**, **latissimus dorsi**, **rhomboids**, **trapezius**, **biceps**, **triceps**, and **serratus anterior** all work together to slowly bring the body down and up.

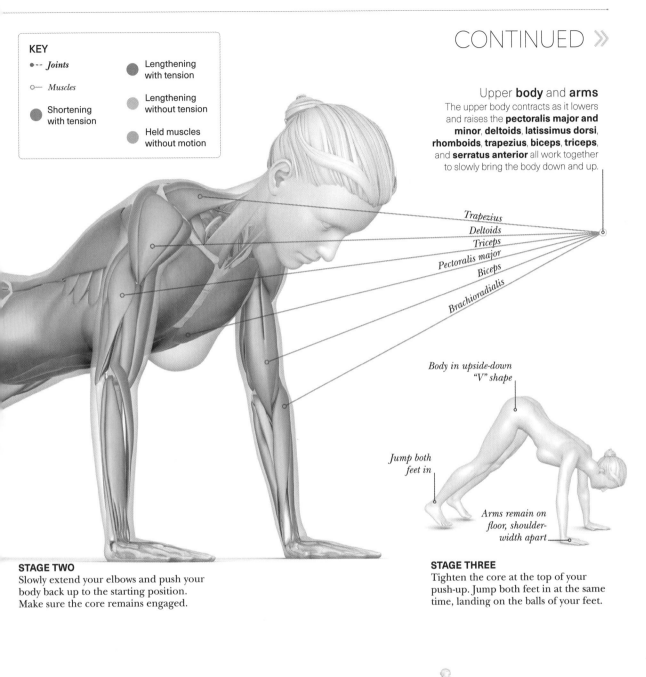

Trapezius
Deltoids
Triceps
Pectoralis major
Biceps
Brachioradialis

Body in upside-down "V" shape

Jump both feet in

Arms remain on floor, shoulder-width apart

STAGE TWO

Slowly extend your elbows and push your body back up to the starting position. Make sure the core remains engaged.

STAGE THREE

Tighten the core at the top of your push-up. Jump both feet in at the same time, landing on the balls of your feet.

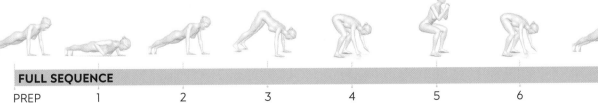

FULL SEQUENCE

PREP   1   2   3   4   5   6   7

» **PUSH-UP** AND **SQUAT**
(CONTINUED)

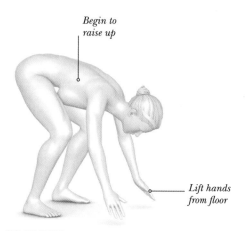

*Begin to
raise up*

*Lift hands
from floor*

STAGE FOUR
As you land, slowly raise your chest
and head, lifting your hands from
the floor and keeping the legs stable.

**THREE-QUARTER
ANTERIOR VIEW**

❝❞

*A squat is a "complex movement"
because of the high amount of
muscle activation. It improves
your **lower body mobility** and
keeps bones and joints healthy.*

STAGE FIVE
Position yourself in a low squat position,
arms in front of you, hands loosely grasped
in front of your chest, and thighs parallel to
the floor. Hold that position for 2–3 seconds.

Upper **body** and **abdominals**

As you squat, your upper body is held in tension. Your **core muscles**, specifically the **erectors**, are activated throughout the movement to stop you from falling forward. They also help brace the spine.

Sternocleidomastoid
Deltoid
Biceps
Triceps
Rectus abdominis

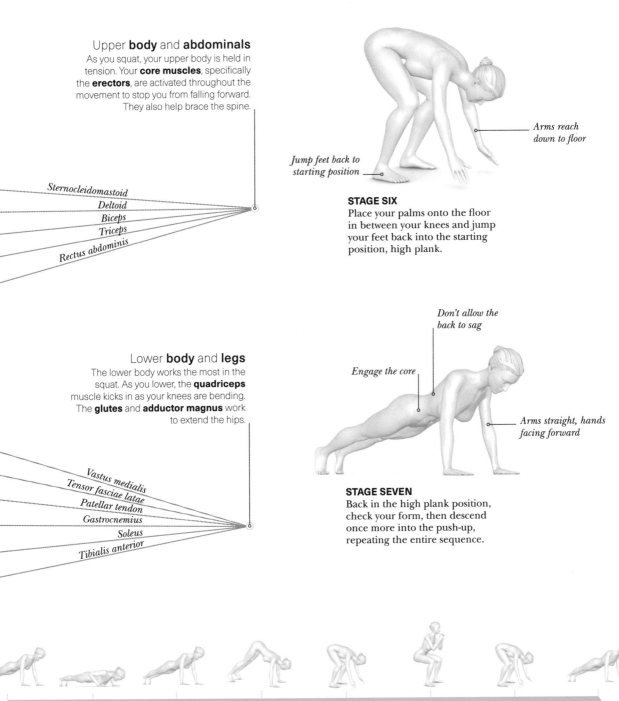

Arms reach down to floor

Jump feet back to starting position

STAGE SIX
Place your palms onto the floor in between your knees and jump your feet back into the starting position, high plank.

Don't allow the back to sag

Engage the core

Arms straight, hands facing forward

STAGE SEVEN
Back in the high plank position, check your form, then descend once more into the push-up, repeating the entire sequence.

Lower **body** and **legs**

The lower body works the most in the squat. As you lower, the **quadriceps** muscle kicks in as your knees are bending. The **glutes** and **adductor magnus** work to extend the hips.

Vastus medialis
Tensor fasciae latae
Patellar tendon
Gastrocnemius
Soleus
Tibialis anterior

FULL SEQUENCE

PREP 1 2 3 4 5 6 7

PUSH-UP AND TUCK JUMP

This is a plyometric full-body workout. The push-up strengthens the chest muscles, shoulders, back of your arms, and "wing" muscles under your armpit. The tuck jump uses your body weight and power to jump up.

THE **BIG PICTURE**

When doing the push-up, remember to keep the abs engaged during the whole movement. It's important to know how to land when doing these plyometric jumps. Make sure you're utilizing the full range of motion—this will help you get farther off the floor. Land softly through the feet, knees, and hips. This will help protect your body from the impact.

Lower **body**
The **quadriceps**, **gluteus maximus and medius**, **hamstrings**, and **calf and shin muscles** are held isometrically to aid in stabilizing the body.

Gluteus maximus
Tensor fasciae latae
Rectus femoris
Vastus lateralis
Tibialis anterior
Peroneus longus

THREE-QUARTER LATERAL VIEW

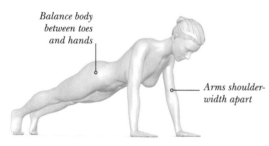

Balance body between toes and hands

Arms shoulder-width apart

PREPARATORY STAGE
Get into a high plank position (see pp.36–37) with your pelvis tucked in, your neck neutral, and your palms directly under your shoulders. Make sure your shoulders are rotated back and down, too, and the core engaged.

STAGE ONE
Inhale, engaging the core. Keeping your back flat, exhale as you slowly lower your body by bending your elbows until your chest grazes the floor. Maintain a straight spine.

! Caution
Make sure you are warmed up before performing this and any plyometric workout. If you're not warmed up, you can strain the knees and joints and cause injuries to the body.

FULL SEQUENCE

| PREP | 1 | 2 | 3 |

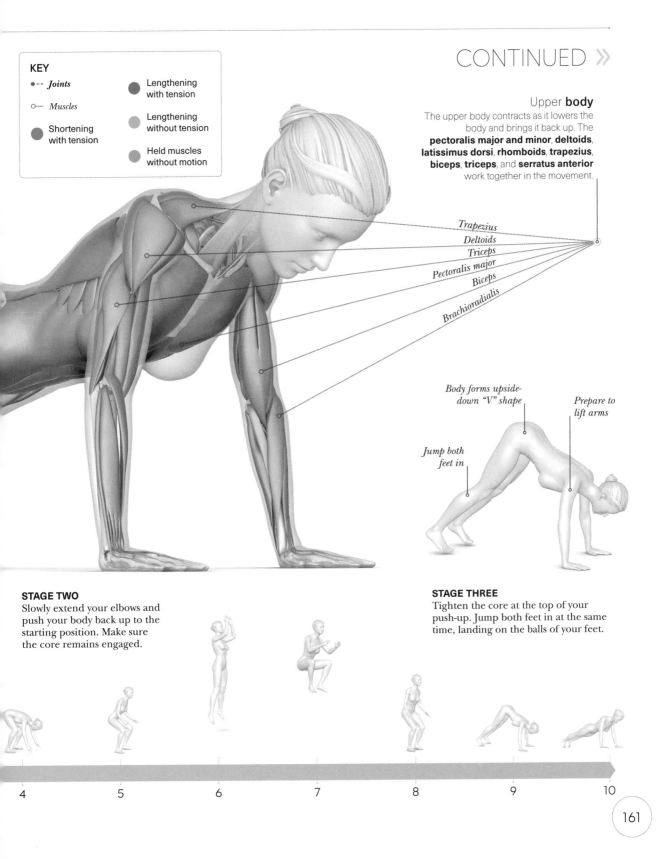

KEY

- •-- *Joints*
- ○— *Muscles*
- ● Shortening with tension
- ● Lengthening with tension
- ● Lengthening without tension
- ● Held muscles without motion

Upper **body**

The upper body contracts as it lowers the body and brings it back up. The **pectoralis major and minor**, **deltoids**, **latissimus dorsi**, **rhomboids**, **trapezius**, **biceps**, **triceps**, and **serratus anterior** work together in the movement.

Trapezius

Deltoids

Triceps

Pectoralis major

Biceps

Brachioradialis

Body forms upside-down "V" shape

Prepare to lift arms

Jump both feet in

STAGE TWO
Slowly extend your elbows and push your body back up to the starting position. Make sure the core remains engaged.

STAGE THREE
Tighten the core at the top of your push-up. Jump both feet in at the same time, landing on the balls of your feet.

4   5   6   7   8   9   10

» **PUSH-UP** AND **TUCK JUMP** (CONTINUED)

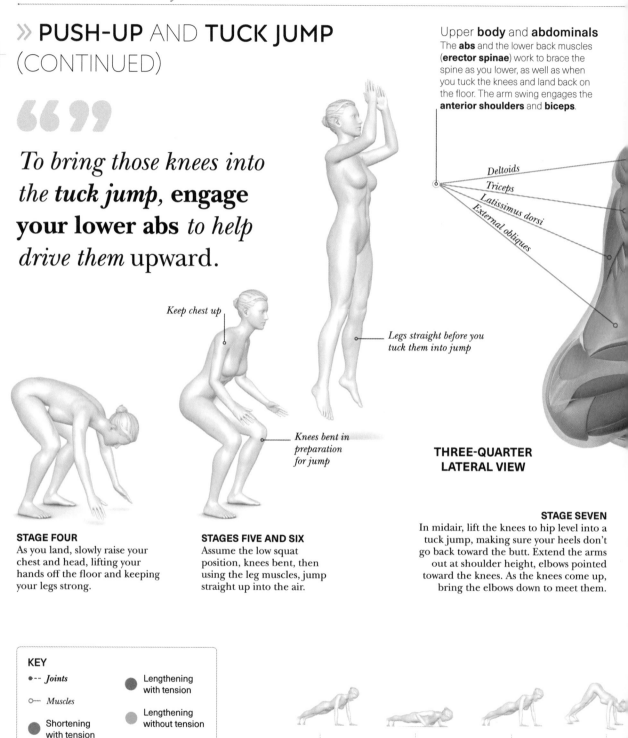

❝❞

*To bring those knees into the **tuck jump**, **engage your lower abs** to help drive them* upward.

Upper **body** and **abdominals**
The **abs** and the lower back muscles (**erector spinae**) work to brace the spine as you lower, as well as when you tuck the knees and land back on the floor. The arm swing engages the **anterior shoulders** and **biceps**.

Deltoids
Triceps
Latissimus dorsi
External obliques

Keep chest up

Legs straight before you tuck them into jump

Knees bent in preparation for jump

THREE-QUARTER LATERAL VIEW

STAGE FOUR
As you land, slowly raise your chest and head, lifting your hands off the floor and keeping your legs strong.

STAGES FIVE AND SIX
Assume the low squat position, knees bent, then using the leg muscles, jump straight up into the air.

STAGE SEVEN
In midair, lift the knees to hip level into a tuck jump, making sure your heels don't go back toward the butt. Extend the arms out at shoulder height, elbows pointed toward the knees. As the knees come up, bring the elbows down to meet them.

KEY

•-- *Joints*

○— *Muscles*

● Shortening with tension

● Lengthening with tension

● Lengthening without tension

● Held muscles without motion

FULL SEQUENCE

PREP   1   2   3

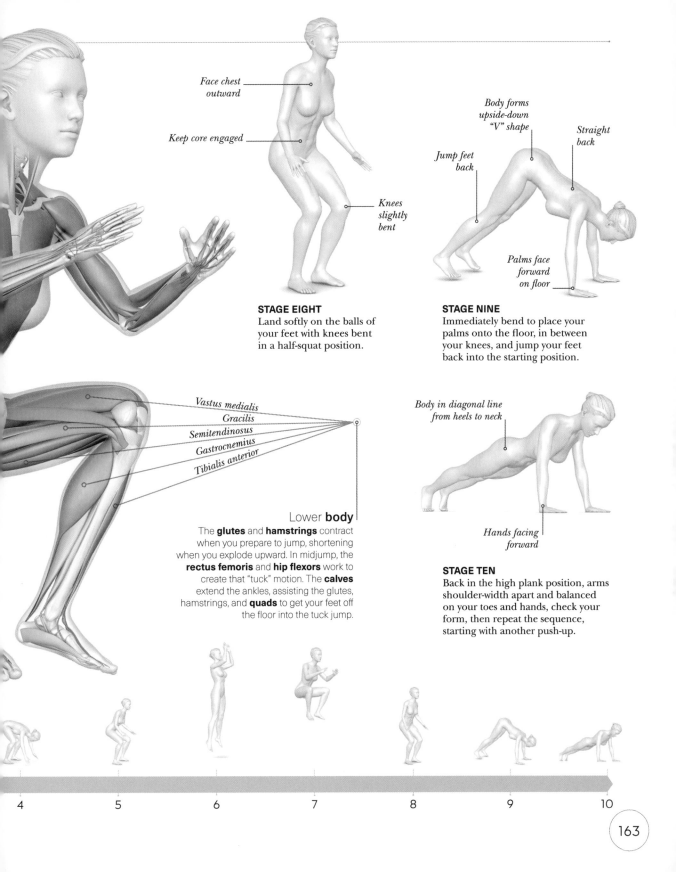

Face chest outward

Keep core engaged

Knees slightly bent

STAGE EIGHT
Land softly on the balls of your feet with knees bent in a half-squat position.

Body forms upside-down "V" shape

Jump feet back

Straight back

Palms face forward on floor

STAGE NINE
Immediately bend to place your palms onto the floor, in between your knees, and jump your feet back into the starting position.

Vastus medialis
Gracilis
Semitendinosus
Gastrocnemius
Tibialis anterior

Lower **body**
The **glutes** and **hamstrings** contract when you prepare to jump, shortening when you explode upward. In midjump, the **rectus femoris** and **hip flexors** work to create that "tuck" motion. The **calves** extend the ankles, assisting the glutes, hamstrings, and **quads** to get your feet off the floor into the tuck jump.

Body in diagonal line from heels to neck

Hands facing forward

STAGE TEN
Back in the high plank position, arms shoulder-width apart and balanced on your toes and hands, check your form, then repeat the sequence, starting with another push-up.

4   5   6   7   8   9   10

BEAR PLANK AND PUSH-UP

This full-body workout focuses on the core and the upper body. The bear plank targets the muscles in your glutes, psoas, quadriceps, shoulders, and arms. The push-up strengthens the chest muscles, shoulders, back of your arms, abdominals, and muscles under your armpit.

Lower **body**
The **quadriceps, gluteus maximus** and **medius, hamstrings,** and **calf and shin muscles** are held isometrically to aid in stabilizing the body.

Gluteus maximus
Tensor fasciae latae
Rectus femoris
Vastus lateralis
Tibialis anterior
Peroneus longus

THE **BIG PICTURE**

When doing the bear plank, try to keep your gaze looking down toward the floor, which will hold your neck in a neutral position. It's an isometric exercise, so stillness is important—try not to shift the hips back and forth. Keep the abs engaged for the entire time of your push-up.

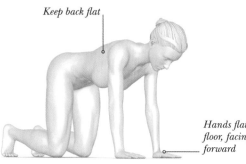

Keep back flat

Hands flat on floor, facing forward

PREPARATORY STAGE
Start on all fours (quadruped [tabletop] position), making sure the back is flat. Hands are shoulder-width apart, wrists are under the shoulders, and knees are hip-width apart. Flex your feet, with your toes on the floor.

 Caution
To avoid strain and pressure on your lower back and joints, make sure your core muscles are engaged, your back is flat, and you have a neutral spine.

Maintain a flat back

Gaze directed at floor

STAGE ONE
Engage your core (belly button to spine), push the palms into the floor, and lift your knees 3–6in (8–15cm) off the floor. Ensure your hips are level with your shoulders. Hold for 30–60 seconds; timing depends on your fitness level.

KEY

- •-- *Joints*
- o— *Muscles*
- ● Shortening with tension
- ● Lengthening with tension
- ● Lengthening without tension
- ● Held muscles without motion

Upper **body**

The upper body contracts as it lowers the body and brings it back up. The **pectoralis major and minor**, **deltoids**, **latissimus dorsi**, **rhomboids**, **trapezius**, **biceps**, **triceps**, and **serratus anterior** all work to raise and lower the body.

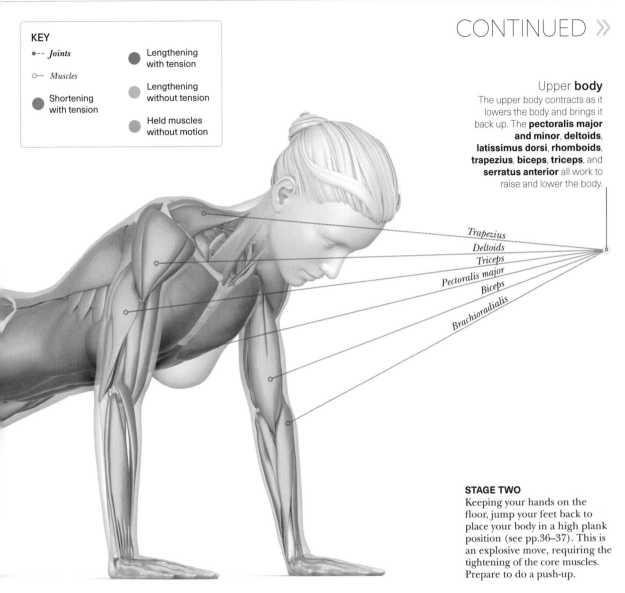

Trapezius
Deltoids
Triceps
Pectoralis major
Biceps
Brachioradialis

THREE-QUARTER LATERAL VIEW

STAGE TWO

Keeping your hands on the floor, jump your feet back to place your body in a high plank position (see pp.36–37). This is an explosive move, requiring the tightening of the core muscles. Prepare to do a push-up.

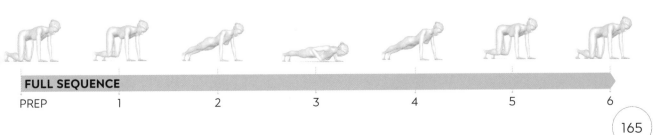

FULL SEQUENCE

PREP   1   2   3   4   5   6

» BEAR PLANK AND PUSH-UP
(CONTINUED)

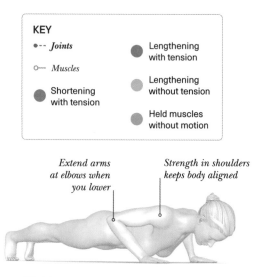

KEY

- •-- *Joints*
- ○— *Muscles*
- ● Shortening with tension
- ● Lengthening with tension
- ● Lengthening without tension
- ● Held muscles without motion

Lower **body**

The muscles in the **hips**, **glutes**, **quadriceps**, and **hamstrings** are engaged during the bear plank hold. The hips and hamstrings, along with the glutes, are further activated as you transition from plank to push-up.

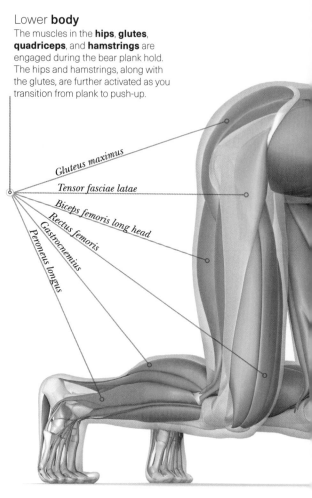

Gluteus maximus

Tensor fasciae latae

Biceps femoris long head

Rectus femoris

Gastrocnemius

Peroneus longus

Extend arms at elbows when you lower

Strength in shoulders keeps body aligned

STAGE THREE
Inhale, pulling the belly in tighter and engaging the core. Exhale and lower your body down by bending your elbows back. Stop when your chest grazes the floor.

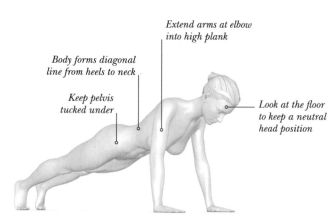

Extend arms at elbow into high plank

Body forms diagonal line from heels to neck

Keep pelvis tucked under

Look at the floor to keep a neutral head position

STAGE FIVE
Slowly and gently jump the feet back into the bear plank position, on all fours, back flat and knees slightly raised from the floor. Hold for 30–60 seconds.

STAGE FOUR
Straighten the arms slowly, and while exhaling, push yourself up back into the high plank position, holding the pose for 2–3 seconds.

*Keeping a **tight core** works as a **stabilizer**, avoiding strain in the wrists.*

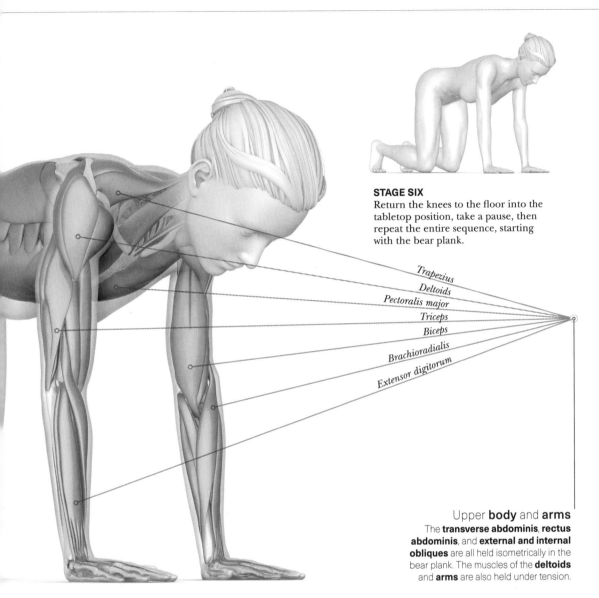

STAGE SIX
Return the knees to the floor into the tabletop position, take a pause, then repeat the entire sequence, starting with the bear plank.

Trapezius
Deltoids
Pectoralis major
Triceps
Biceps
Brachioradialis
Extensor digitorum

Upper **body** and **arms**
The **transverse abdominis**, **rectus abdominis**, and **external and internal obliques** are all held isometrically in the bear plank. The muscles of the **deltoids** and **arms** are also held under tension.

THREE-QUARTER LATERAL VIEW

FULL SEQUENCE

PREP   1   2   3   4   5   6

HIGH PLANK, ANKLE TAP, AND PUSH-UP

This exercise sequence improves balance, coordination, and posture and strengthens your core. The transition from each stage improves body flexibility and makes the tummy tight. The alternating of the ankle tap further engages the obliques.

THE **BIG PICTURE**

You need balance and coordination to complete this exercise correctly, as it has several components. Remember to engage your core throughout the three exercises—the high plank, ankle tap, and push-up—and keep your legs strong, as they act as stabilizers to help prevent any sagging or arching of the spinal column.

KEY

- •-- *Joints*
- ○- *Muscles*
- ● Shortening with tension
- ● Lengthening with tension
- ● Lengthening without tension
- ● Held muscles without motion

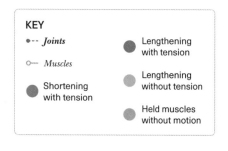

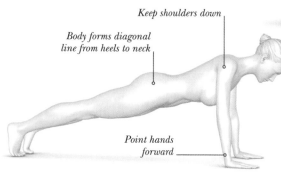

Keep shoulders down

Body forms diagonal line from heels to neck

Point hands forward

PREPARATORY STAGE
Start in the high plank position (see pp.36–37), arms shoulder-width apart, back straight, and weight resting between your toes and your hands on the floor. Engage the core.

STAGE ONE
Raise your body into an inverted "V" shape by bringing your hips into the air. Simultaneously raise your left hand off the floor and bring it back toward the right ankle, holding the position for 2–3 seconds. Your head will naturally rotate to the right as you do this.

! Caution
Make sure that you maintain a neutral spine and that your shoulders are down, not creeping up toward your ears. Keeping the abs engaged during the push-up will stop the spine from caving in and causing pressure on the lower back and joints.

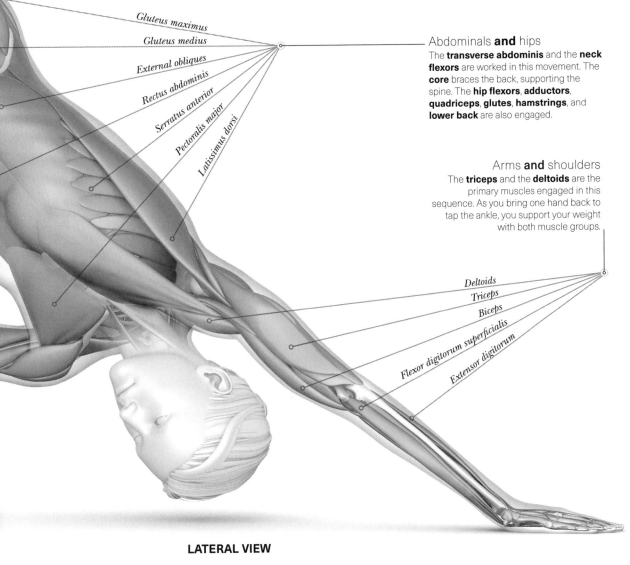

Gluteus maximus

Gluteus medius

External obliques

Rectus abdominis

Serratus anterior

Pectoralis major

Latissimus dorsi

Abdominals **and** hips

The **transverse abdominis** and the **neck flexors** are worked in this movement. The **core** braces the back, supporting the spine. The **hip flexors**, **adductors**, **quadriceps**, **glutes**, **hamstrings**, and **lower back** are also engaged.

Arms **and** shoulders

The **triceps** and the **deltoids** are the primary muscles engaged in this sequence. As you bring one hand back to tap the ankle, you support your weight with both muscle groups.

Deltoids

Triceps

Biceps

Flexor digitorum superficialis

Extensor digitorum

LATERAL VIEW

FULL SEQUENCE

PREP   1   2   3   4   5

» HIGH PLANK, ANKLE TAP, AND PUSH-UP
(CONTINUED)

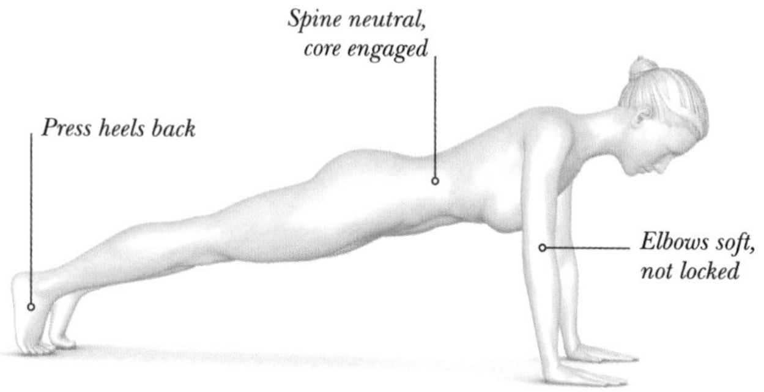

Maintain a flat back

Reach right hand toward the left ankle

Left hand provides support on the floor

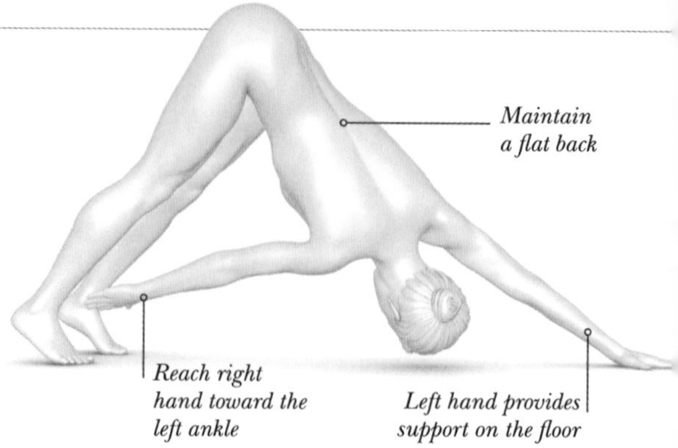

STAGE THREE
Raise up into the inverted "V" again and repeat the ankle tap on the other side, your right hand reaching toward the left ankle.

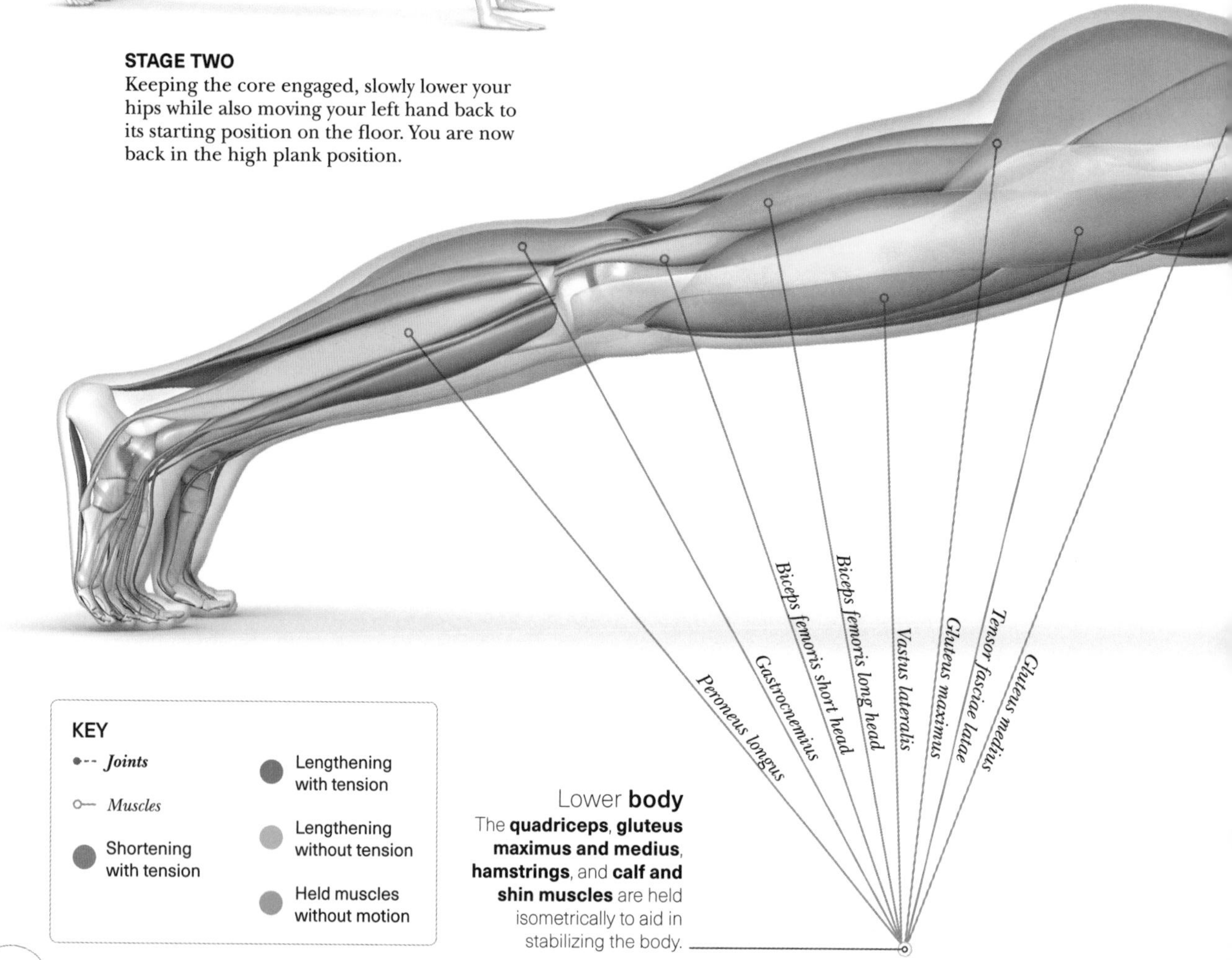

Spine neutral, core engaged

Press heels back

Elbows soft, not locked

STAGE TWO
Keeping the core engaged, slowly lower your hips while also moving your left hand back to its starting position on the floor. You are now back in the high plank position.

Biceps femoris short head

Biceps femoris long head

Vastus lateralis

Gluteus maximus

Tensor fasciae latae

Gluteus medius

Peroneus longus

Gastrocnemius

KEY

•-- *Joints*

o— *Muscles*

● Shortening with tension

● Lengthening with tension

● Lengthening without tension

● Held muscles without motion

Lower **body**
The **quadriceps**, **gluteus maximus and medius**, **hamstrings**, and **calf and shin muscles** are held isometrically to aid in stabilizing the body.

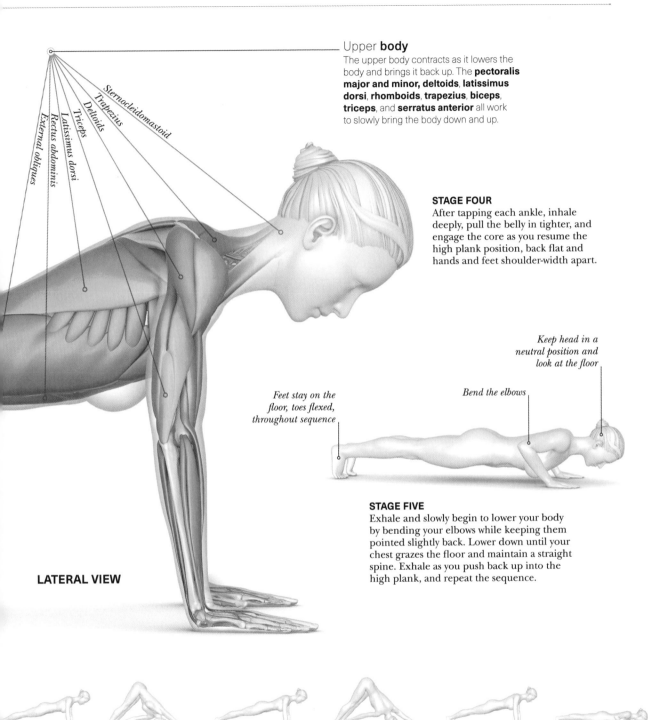

Upper **body**

The upper body contracts as it lowers the body and brings it back up. The **pectoralis major and minor, deltoids, latissimus dorsi, rhomboids, trapezius, biceps, triceps,** and **serratus anterior** all work to slowly bring the body down and up.

Sternocleidomastoid

Trapezius

Deltoids

Triceps

Latissimus dorsi

Rectus abdominis

External obliques

STAGE FOUR

After tapping each ankle, inhale deeply, pull the belly in tighter, and engage the core as you resume the high plank position, back flat and hands and feet shoulder-width apart.

Keep head in a neutral position and look at the floor

Bend the elbows

Feet stay on the floor, toes flexed, throughout sequence

STAGE FIVE

Exhale and slowly begin to lower your body by bending your elbows while keeping them pointed slightly back. Lower down until your chest grazes the floor and maintain a straight spine. Exhale as you push back up into the high plank, and repeat the sequence.

LATERAL VIEW

FULL SEQUENCE

PREP   1   2   3   4   5

B BOY
POWER KICKS

This core-focused workout strengthens the obliques, abs, and lower back. It also strengthens your shoulders, arms, and legs and improves your cardiovascular strength and endurance.

THE **BIG PICTURE**

This is an advanced cardiovascular movement. If you are a beginner, start off moving at a slower pace so that you can perfect your form. Begin by performing the exercise for 30-second intervals, aiming for 3–5 sets. As you get stronger and more accustomed to the move, you can increase both the speed and time of the exercise.

Right leg stays on floor

Left arm takes the weight as you kick through

Left leg comes under body to kick

STAGE ONE
Exhale and lift your right hand and left foot off the floor, rotate your hips to the right, and place your right heel on the floor as you kick your left leg to the right, underneath your body. Extend your left leg and touch that heel briefly on the floor, rotating your body so that it is almost facing the ceiling. Bring your right arm above your head.

PREPARATORY STAGE
Get down on your hands and knees and lift your knees 3–6in (8–15cm) off the floor, into a bear plank position (see pp.44–45). Position your wrists under your shoulders, your knees in line with the hips, and your back flat.

Lower **body**
When you are kicking the leg through in this exercise, the muscles in the **quadriceps**, **hamstrings**, and **glutes** are all held isometrically. The leg slide-through engages the quadriceps; the glutes stabilize the hips.

Sartorius
Gastrocnemius
Tibialis anterior
Rectus femoris
Vastus lateralis
Biceps femoris longus
Gluteus maximus

ANTERIOR-LATERAL VIEW

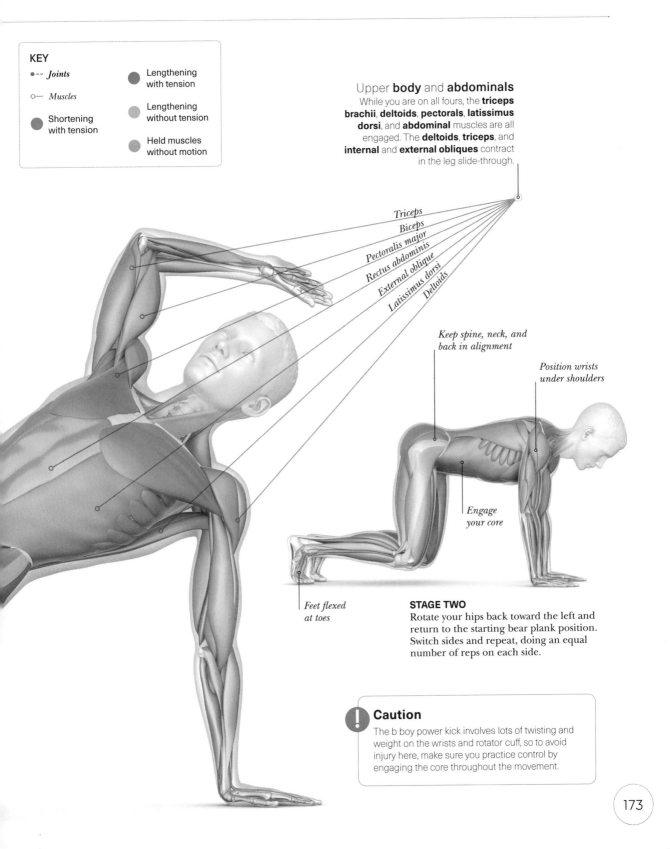

Upper **body** and **abdominals**

While you are on all fours, the **triceps brachii, deltoids, pectorals, latissimus dorsi**, and **abdominal** muscles are all engaged. The **deltoids, triceps**, and **internal** and **external obliques** contract in the leg slide-through.

Triceps

Biceps

Pectoralis major

Rectus abdominis

External oblique

Latissimus dorsi

Deltoids

Keep spine, neck, and back in alignment

Position wrists under shoulders

Engage your core

Feet flexed at toes

STAGE TWO

Rotate your hips back toward the left and return to the starting bear plank position. Switch sides and repeat, doing an equal number of reps on each side.

! Caution

The b boy power kick involves lots of twisting and weight on the wrists and rotator cuff, so to avoid injury here, make sure you practice control by engaging the core throughout the movement.

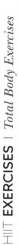

MILITARY PRESS AND OVERHEAD TRICEPS EXTENSION

The military press movement strengthens the chest, shoulders, arms, and upper back muscles, as well as the muscles in your core. The triceps extension isolates and strengthens the triceps muscles.

THE BIG PICTURE

Use an overhand grip on the dumbbells, fingers facing out. Make sure you choose a weight that allows you to maintain the correct form. It's important to watch your elbows—keep them either directly underneath your wrists or slightly more inward. Beginners can start with 3 sets of 8–10 reps, progressing the weight as you become accustomed to the exercise.

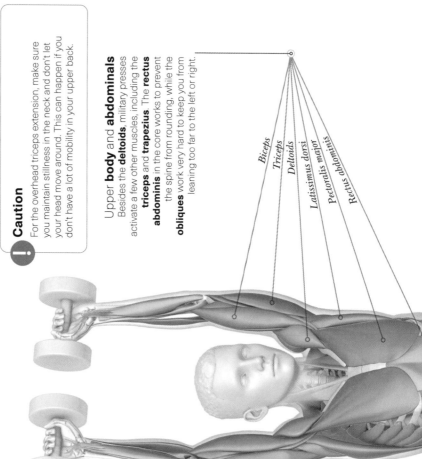

Biceps
Triceps
Deltoids
Latissimus dorsi
Pectoralis major
Rectus abdominis

Caution

For the overhead triceps extension, make sure you maintain stillness in the neck and don't let your head move around. This can happen if you don't have a lot of mobility in your upper back.

Upper **body** and **abdominals**
Besides the **deltoids**, military presses activate a few other muscles, including the **triceps** and **trapezius**. The **rectus abdominis** in the core works to prevent the spine from rounding, while the **obliques** work very hard to keep you from leaning too far to the left or right.

CONTINUED »

KEY

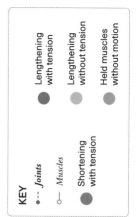

- ·--- *Joints*
- o— *Muscles*
- ● Shortening with tension
- ● Lengthening with tension
- ● Lengthening without tension
- ● Held muscles without motion

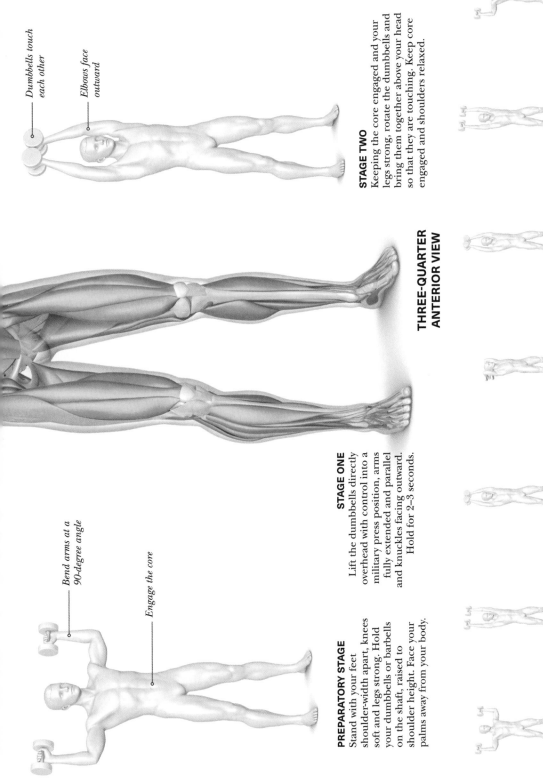

Dumbbells touch each other

Elbows face outward

STAGE TWO
Keeping the core engaged and your legs strong, rotate the dumbbells and bring them together above your head so that they are touching. Keep core engaged and shoulders relaxed.

THREE-QUARTER ANTERIOR VIEW

STAGE ONE
Lift the dumbbells directly overhead with control into a military press position, arms fully extended and parallel and knuckles facing outward. Hold for 2–3 seconds.

Bend arms at a 90-degree angle

Engage the core

PREPARATORY STAGE
Stand with your feet shoulder-width apart, knees soft and legs strong. Hold your dumbbells or barbells on the shaft, raised to shoulder height. Face your palms away from your body.

FULL SEQUENCE

PREP   1   2   3   4   5   6

175

» MILITARY PRESS AND OVERHEAD TRICEPS EXTENSION (CONTINUED)

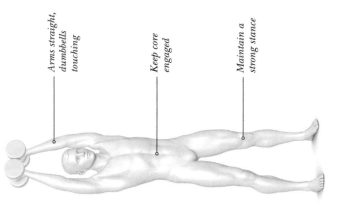

Arms straight, dumbbells touching

Keep core engaged

Maintain a strong stance

STAGE FOUR
Once you reach a 90-degree bend at the elbow or slightly farther, inhale and reverse the movement, bringing the weights back above your head, still facing each other.

Upper **body** and **abdominals**
The **triceps** are isolated in this move, the three heads working together to extend the forearm at the elbow joint. The **core** remains engaged throughout.

Triceps

Deltoids

Latissimus dorsi

Pectoralis major

Rectus abdominis

STAGE THREE
Exhale and bend the elbows to slowly lower the weights down, bringing them behind the head into a triceps extension. The weights should not touch the back of the head when they are in their lowest position. Keep the chest aligned over the hips and don't arch the back.

You want to isolate the triceps extension movement only at the elbow joint—keep everything else engaged and still.

Don't let elbows flare out

Engage core to provide stability

Maintain a straight spine

Elbows should be parallel to shoulders

Maintain a straight back

Keep your knees soft but strong

STAGE FIVE

Separate and rotate the weights back to the top of the military press position, the arms straight, holding the weights above you, palms facing outward and knuckles facing upward.

STAGE SIX

Bend the elbows to lower the weights slowly until your elbows are back at shoulder height, in the preparatory stage position. Repeat the entire sequence.

THREE-QUARTER ANTERIOR VIEW

KEY

- •-- Joints
- o— Muscles
- ⬤ Shortening with tension
- ⬤ Lengthening with tension
- ⬤ Lengthening without tension
- ⬤ Held muscles without motion

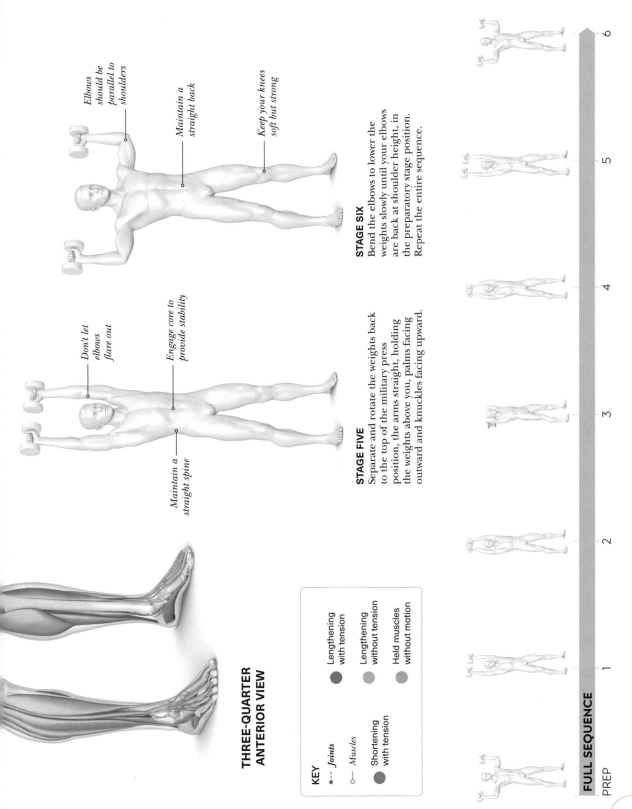

FULL SEQUENCE

PREP | 1 | 2 | 3 | 4 | 5 | 6

BENT-OVER ROW AND HAMMER CURL

POSTERIOR-LATERAL VIEW

This is a compound functional exercise, aimed at strengthening several muscles. The row strengthens the back, chest, upper arm, and rotator cuff muscles. The hammer curl targets the biceps muscles.

THE **BIG PICTURE**

When bending over to perform the double row move, don't bend more than 45 degrees and keep the back straight, not curved, and the shoulders square throughout. In the hammer curl, lift the weights slowly and with control—don't swing them—keeping your elbows in a stable, fixed position. Count to 3 on the way up and 3 on the way down. Complete 3 sets of 8 reps of the entire sequence.

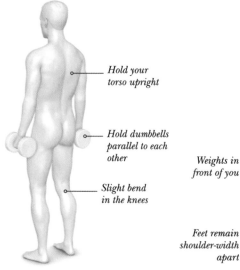

Hold your torso upright

Hold dumbbells parallel to each other

Slight bend in the knees

Bend back at a 45-degree angle

Weights in front of you

Feet remain shoulder-width apart

PREPARATORY STAGE
Start by standing with feet shoulder-width apart, knees slightly bent. Hold a dumbbell in each hand, arms by your sides.

STAGE ONE
Inhale and bend at the hip at about a 45-degree angle, bending the knees and keeping the back straight. Your weights will now be in front of you.

STAGE TWO
Exhale and lift the weights up into a reverse row, with arms bent at the elbows to 90 degrees. While lifting, the arms should only go to a point just below your shoulders. Keep the chest slightly raised.

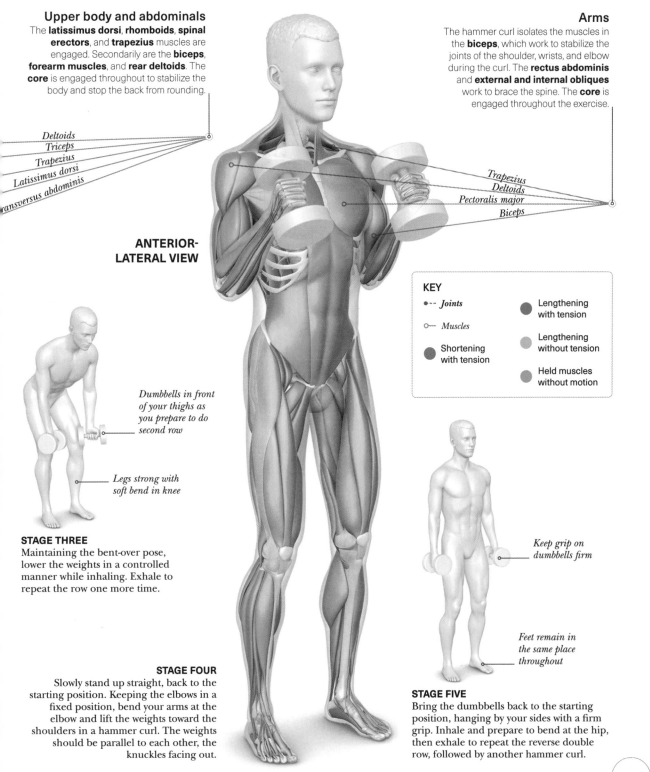

Upper body and abdominals

The **latissimus dorsi**, **rhomboids**, **spinal erectors**, and **trapezius** muscles are engaged. Secondarily are the **biceps**, **forearm muscles**, and **rear deltoids**. The **core** is engaged throughout to stabilize the body and stop the back from rounding.

Arms

The hammer curl isolates the muscles in the **biceps**, which work to stabilize the joints of the shoulder, wrists, and elbow during the curl. The **rectus abdominis** and **external and internal obliques** work to brace the spine. The **core** is engaged throughout the exercise.

Deltoids
Triceps
Trapezius
Latissimus dorsi
Transversus abdominis

Trapezius
Deltoids
Pectoralis major
Biceps

ANTERIOR-LATERAL VIEW

KEY

•-- *Joints*

○— *Muscles*

⬤ Shortening with tension

⬤ Lengthening with tension

⬤ Lengthening without tension

⬤ Held muscles without motion

Dumbbells in front of your thighs as you prepare to do second row

Legs strong with soft bend in knee

STAGE THREE

Maintaining the bent-over pose, lower the weights in a controlled manner while inhaling. Exhale to repeat the row one more time.

STAGE FOUR

Slowly stand up straight, back to the starting position. Keeping the elbows in a fixed position, bend your arms at the elbow and lift the weights toward the shoulders in a hammer curl. The weights should be parallel to each other, the knuckles facing out.

Keep grip on dumbbells firm

Feet remain in the same place throughout

STAGE FIVE

Bring the dumbbells back to the starting position, hanging by your sides with a firm grip. Inhale and prepare to bend at the hip, then exhale to repeat the reverse double row, followed by another hammer curl.

179

REAR **DELTOID FLY** AND **TRICEPS** KICKBACK

This compound exercise works several muscle groups at once. It targets the deltoids at the back of the shoulders and the major muscles of the upper back, including the trapezius, as well as working the triceps and abdominals.

THE **BIG PICTURE**

You will begin with the rear deltoid fly, in which scapular retraction occurs, causing the shoulder blades to pull in toward each other. Remaining in the bent-over position, you then perform the triceps kickback. Choose a weight that is appropriate for your fitness level. Practice without weight first until you get comfortable with the movement. Try 1–3 sets of 8–12 reps to start, slowly increasing weight as you get stronger.

POSTERIOR-LATERAL VIEW

Spine in neutral position

KEY

•-- *Joints*

○— *Muscles*

● Shortening with tension

● Lengthening with tension

● Lengthening without tension

● Held muscles without motion

Dumbbells hang straight down with palms facing you

Legs shoulder-width apart, slight bend in knees

PREPARATORY STAGE
Stand with feet shoulder-width apart holding the dumbbells hanging down in front of you. Press the hips back in a hinge motion, bringing your chest forward.

STAGE ONE

Exhale and raise both arms to the side, squeezing the shoulder blades together. Keep a soft bend in your elbows as you pull your shoulder blades toward the spine. Try to hold for 2 seconds.

CONTINUED »

Extensor digitorum
Brachioradialis
Triceps
Trapezius
Infraspinatus
Triceps
Serratus anterior
Iliocostalis
Transversus abdominis

Engage core throughout movement

Return weights to starting position, hanging in front of you

Keep knees soft

Upper **body**

The reverse fly targets several muscles in the **upper back** and **deltoids**. The **posterior deltoids** are the main muscles recruited, in addition to the **middle and lower trapezius**, **rhomboids**, **infraspinatus**, and **teres minor**.

STAGE TWO

Inhale as you lower the weight back to the starting position. Avoid rounding your back and hunching the shoulders forward. Tuck your chin to maintain a neutral spine and breathe normally.

! Caution

Avoid rounding the back, as that puts strain on the spine. When the weights are too heavy, this can lead to rounding the back and swinging of the weights, which uses momentum to lift the weights instead of using the intended muscle.

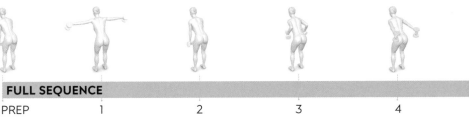

FULL SEQUENCE

PREP   1   2   3   4   5   6

» REAR **DELTOID FLY** AND **TRICEPS** KICKBACK (CONTINUED)

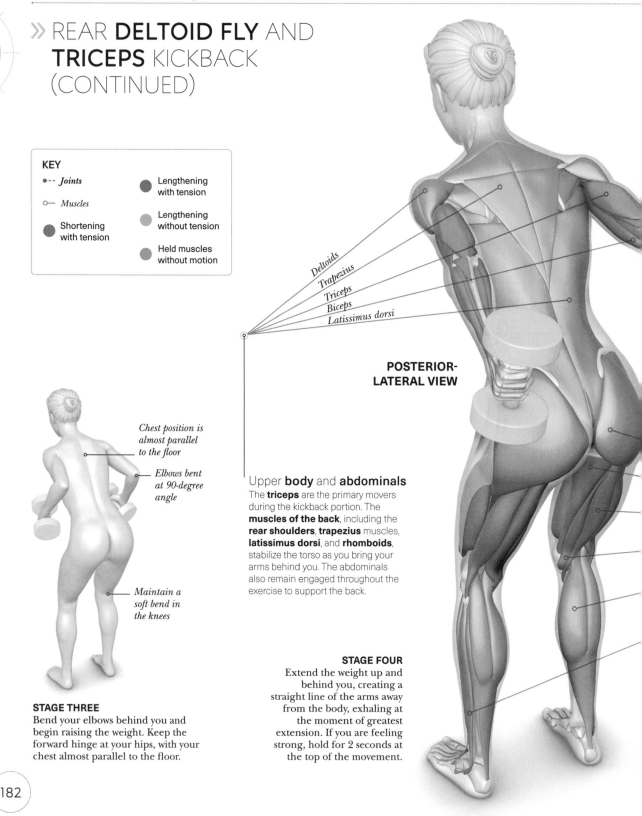

KEY

•-- *Joints*

o— *Muscles*

● Shortening with tension

● Lengthening with tension

○ Lengthening without tension

◐ Held muscles without motion

Deltoids

Trapezius

Triceps

Biceps

Latissimus dorsi

POSTERIOR-LATERAL VIEW

Chest position is almost parallel to the floor

Elbows bent at 90-degree angle

Maintain a soft bend in the knees

Upper **body** and **abdominals**

The **triceps** are the primary movers during the kickback portion. The **muscles of the back**, including the **rear shoulders**, **trapezius** muscles, **latissimus dorsi**, and **rhomboids**, stabilize the torso as you bring your arms behind you. The abdominals also remain engaged throughout the exercise to support the back.

STAGE THREE

Bend your elbows behind you and begin raising the weight. Keep the forward hinge at your hips, with your chest almost parallel to the floor.

STAGE FOUR

Extend the weight up and behind you, creating a straight line of the arms away from the body, exhaling at the moment of greatest extension. If you are feeling strong, hold for 2 seconds at the top of the movement.

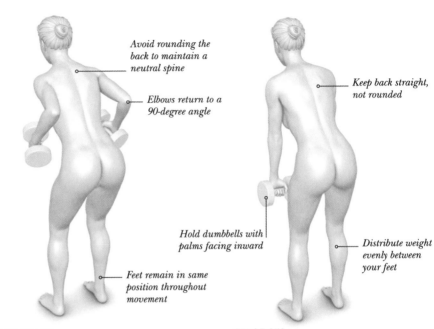

Avoid rounding the back to maintain a neutral spine

Elbows return to a 90-degree angle

Feet remain in same position throughout movement

Keep back straight, not rounded

Hold dumbbells with palms facing inward

Distribute weight evenly between your feet

STAGE FIVE

Inhale as you slowly bring the weights back toward the body. Don't swing the weights: stop them for a split second to reduce momentum.

STAGE SIX

Straighten the arms and drop the dumbbells back to the preparatory stage position. Reset, then perform the reverse deltoid fly again to complete one rep.

Gluteus maximus
Adductor magnus
Semitendinosus
Semimembranosus
Gastrocnemius
Peroneus longus

Lower body

The lower body works as a stabilizer in this exercise. The **glutes** are held isometrically throughout. Squeezing the glutes holds the hips in place and works to keep the spine neutral.

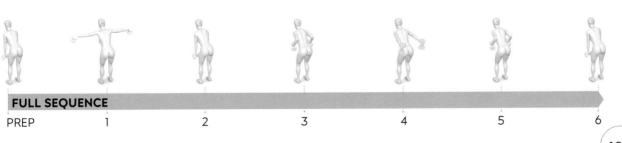

FULL SEQUENCE

PREP   1   2   3   4   5   6

183

SUMO SQUAT AND HAMMER CONCENTRATION CURL

This sequence strengthens the glutes, quads, hamstrings, hip flexors, calves, and core, placing emphasis on the hips, as well as the inner thighs. The hammer concentration curls strengthen the biceps muscles at the front of the upper arm, and also the muscles of the lower arm—the brachialis and brachioradialis.

THE **BIG PICTURE**

In the sumo squat, don't let your knees cave inward. Keeping your chest upright and not hunched throughout the sequence and your core engaged will help you maintain proper form. Elbows stay glued to your thigh during the concentration curl.

KEY

•-- *Joints*

○— *Muscles*

● Shortening with tension

● Lengthening with tension

● Lengthening without tension

● Held muscles without motion

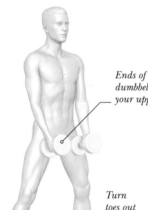

Ends of dumbbells touch your upper thigh

Turn toes out

Maintain an upright chest

Dumbbells between your legs, parallel to each other

THREE-QUARTER ANTERIOR VIEW

PREPARATORY STAGE
Stand with your legs far apart and turn your toes out 45 degrees. Hold the dumbbells in front of your thighs, arms relaxed, and prepare to squat.

STAGE ONE
Start by bending at the hips and knees and slowly push your hips back. Keep the chest up and knees out as you lower into the squat. Bring your dumbbells between your legs.

STAGE TWO
Holding the sumo squat, place the elbows on the top of the thighs, palms facing the midline of the body. Hinge at the elbow and lift both dumbbells up until they are touching the shoulders.

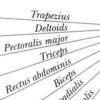

Arms touch the inner thighs

Hold weights parallel to each other

STAGE THREE
Maintaining the sumo squat position, lower the weights to their starting position, between your legs, chest facing forward.

Upper **body and abdominals**
The **arms** lengthen under tension while holding the dumbbells. The spine is stabilized by the **abdominal muscles**.

Trapezius
Deltoids
Pectoralis major
Triceps
Rectus abdominis
Biceps
Brachioradialis
Flexor digitorum superficialis

Trapezius
Deltoids
Pectoralis major
Biceps

Upper **body**
The **biceps brachii, brachialis, triceps brachii, flexor digitorum, pectorals**, and **serratus anterior** are all targeted. The **rectus abdominis** braces the spine.

Vastus medialis
Rectus femoris
Tibialis anterior
Gastrocnemius
Soleus

Lower **body**
The sumo squat strengthens the **quadriceps**, **glutes** muscles, **hips**, **hamstrings**, and **calves**, with a specific focus on the **inner thighs and abductors**.

Vastus lateralis
Vastus medialis
Rectus femoris
Tibialis anterior
Gastrocnemius
Soleus

THREE-QUARTER ANTERIOR VIEW

Lower **body**
The muscles in the **hamstrings, quadriceps, adductors, abductors**, and **glutes** are held isometrically at the bottom of the sumo squat.

STAGE FOUR
Straighten the legs and slowly stand back up to resume the starting position, dumbbells pressed against your upper thighs. Reset and prepare to repeat stages 2 and 3.

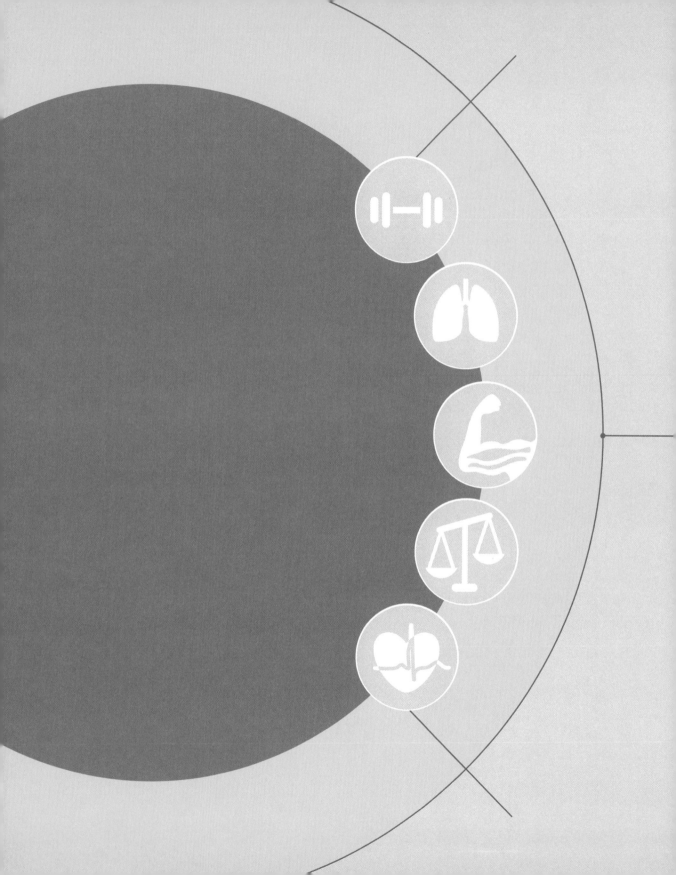

HIIT
TRAINING

This section contains 42 HIIT workout routines aimed at beginner, intermediate, and advanced levels and focusing on total body, upper body, or lower body. You perform each exercise for a certain amount of time based on your fitness and complete 5 rounds. Routines can also be stacked for a longer workout. You will also find advice on warming up and cooling down, how to plan your training, and how to create your own workouts.

GETTING STARTED

Before performing any exercises, find the workouts that are right for you. There are training plans in this book designed for beginner, intermediate, and advanced users. To get the most out of them, it's important to start where you are and build up the muscular and cardiovascular strength to progress. Use the fitness test below to determine where you should begin.

" "

*Before you get into HIIT, you need to assess your **current fitness** level. This simple assessment will let you know where to start and gives a **baseline for measuring** your progress.*

ASSESS YOUR FITNESS

Before beginning this program, use this bodyweight resistance test to determine your current fitness level. The results will indicate where you are in your fitness journey.

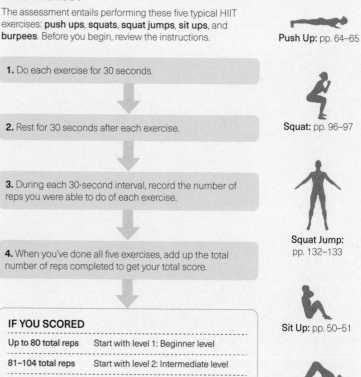

Take the test

The assessment entails performing these five typical HIIT exercises: **push ups**, **squats**, **squat jumps**, **sit ups**, and **burpees**. Before you begin, review the instructions.

1. Do each exercise for 30 seconds.

2. Rest for 30 seconds after each exercise.

3. During each 30-second interval, record the number of reps you were able to do of each exercise.

4. When you've done all five exercises, add up the total number of reps completed to get your total score.

IF YOU SCORED

Up to 80 total reps	Start with level 1: Beginner level
81–104 total reps	Start with level 2: Intermediate level
105 or higher	Start with level 3: Advanced level

Push Up: pp. 64–65

Squat: pp. 96–97

Squat Jump: pp. 132–133

Sit Up: pp. 50–51

Burpee: pp. 146–149

WHAT'S THE RIGHT LEVEL FOR YOU?

After completing the fitness assessment, identify whether you are a beginner, intermediate, or advanced user. Start your training at that level.

Level 1
Beginner level: for those new to HIIT

If you scored up to 80 reps, you are starting at the beginner level. Starting here creates a good solid foundation, so you can eventually progress to more advanced HIIT workouts. Use this time to focus on your form and proper breathing techniques. Start with just bodyweight and little to no weight.

Duration per exercise: 30 seconds
Breaks between exercises: 15 seconds
Number of sets: 2–3
Duration of rest between sets: 30–60 seconds

Level 2
Intermediate

Starting here means you currently have a strong foundation. However, it is still always better to ease into this type of workout with lighter resistance. You can challenge yourself here with longer rep times and shorter rest times, and then with heavier weights as your body becomes conditioned.

Duration per exercise: 45 seconds
Breaks between exercises: 15 seconds
Number of sets: 3–4
Duration of rest between sets: 30–45 seconds

Level 3
Advanced

A score of 105+ means you are already well conditioned, but it may be time to push yourself and take it to the next level. Improve upon your cardiovascular and muscular endurance and strength by challenging yourself with heavier weights, shorter rest times, and longer durations of work time.

Duration per exercise: 60 seconds
Breaks between exercises: None
Number of sets: 4–5
Duration of rest between sets: 30–45 seconds

BODY FAT COMPOSITION

HIIT is known to burn fat even after your workout is complete (see pp. 16–17). Before you begin this journey, it may be advantageous to start by seeing where your body currently sits. Body fat composition describes different components that make up total body weight: muscles, bones, and fat. There are several ways to calculate body fat: through body mass index (BMI); skinfold measurements using calipers; or with a tape measure. There are also several online body fat calculators you can use to help you.

CALCULATING YOUR BODY FAT

Use the following calculation for a broad assessment of your body fat percentage determined by BMI.

FORMULAE
Standard: weight (lb) ÷ height^2 (in) x 703
Metric: weight (kg) ÷ height^2 (m)

Examples
160 lb ÷ 65 in^2 x 703 = BMI 26.62 (overweight)
65 kg ÷ 1.8 m^2 = BMI 20.06 (normal)

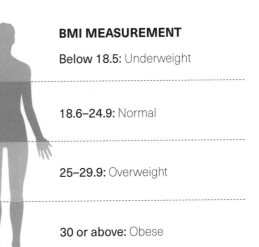

BMI MEASUREMENT

Below 18.5: Underweight

18.6–24.9: Normal

25–29.9: Overweight

30 or above: Obese

PLANNING YOUR TRAINING

When I was preparing for bodybuilding competitions, I planned everything: my workout, meal plan, and diet for 13 weeks. A clear plan creates a pathway to our goals. Training is important, but beware of overtraining, and use proper recovery techniques, along with a nutrition plan geared towards fueling your body.

HOW OFTEN SHOULD I TRAIN?

HIIT training isn't meant to be done every day. These super-quick routines were designed so you could perform them at maximum effort. We all want to get in shape and lose weight fast, so it's no wonder that HIIT training is such a good idea. It's a shorter exercise aimed at maximizing the heart rate, to boost the metabolism and burn fat. HIIT workouts eliminate the need to spend exhausting hours on the treadmill. The fat-burning phenomenon that occurs after a HIIT workout is known as EPOC (excess post-exercise oxygen consumption, see pp. 16–17)), which equates to more calories burned, a boost to the metabolism, and burning fat.

Because of the intensity involved in HIIT training, it's best to ease into it. The fitness assessment test (see p. 188) will tell you which level to start with. It's recommended you start training three to four times a week. If you are new to HIIT, perhaps start with once a week, until your body is conditioned. As your body gets used to the workouts, you can increase the number of days, but give yourself 24 hours in between sessions.

Overtraining your muscles is one reason to avoid training too much. When you do too much physical activity, in combination with too little rest and no recovery periods, it can place stress on joints and muscles. And when your muscles are sore and overworked, this can affect your physical performance and lead to injuries.

Weekly program

When planning your weekly workouts, it's important to incorporate different muscle groups. For instance, you don't want to do back-to-back lower body workouts and risk overworking your leg muscles. It's also vital you space your sessions out, to allow for proper recovery in between.

FOR EXAMPLE

If you choose to work out four times a week, start with an upper body focused workout on day one, a lower body workout on day two, abs-focused workout on day three, and full-body workout on day four.

Training for your level

After taking the fitness assessment test, you will have a better understanding of where you should start your fitness journey.

Level 1: Beginner

If you're starting here, ease into HIIT training with just one or two training days a week. In addition, make sure you stick to the number of sets, duration of exercises, and duration of rests recommended in the fitness assessment test.

Level 2: Intermediate

If the fitness assessment test identifies you as an intermediate user, start HIIT training with two to three training days a week. In addition, make sure you stick to the number of sets, duration of exercises, and duration of rests recommended in the fitness assessment test.

Level 3: Advanced

If you are starting here, aim to have three to four days training a week. In addition, make sure you stick to the number of sets, duration of exercises, and duration of rests recommended in the assessment test.

HOW DO YOU **PROGRESS** YOUR **TRAINING?**

One of my favourite sayings is "you can always take it up a notch." If you feel you are plateauing, there are several ways you can intensify your workout. Here are your options.

Adding reps and load

Increasing the duration of your exercises, in addition to the amount of resistance you use, is a surefire way of building muscle size (hypertrophy). As a consequence, you'll also find the workouts more challenging.

Adding sets

One way of taking your workouts to the next level is to increase the number of sets, which means you'll also be spending more time doing the series of exercises. Increasing the number of sets like this will certainly push your body.

Progressive reps in reserve (RIR)

Reps in reserve (RIR) is the number of reps you have left in the "tank" after completing a set—in other words, how many more reps you could have done before reaching failure on a set. Aim to train to within 4–5 reps of failure to maximize stimulus.

Recovery

As important as the workout—if not more important—is recovery. It may be that you are overtraining and not giving your body the opportunity it needs to properly recover and lock in the gains to be had from training. There are several recovery tools you can use. I go into more detail below.

RECOVERY

Another of my favourite sayings is "you should recover just as hard as you train." Over the age of 30, you should be recovering at a 1:1 ratio; so, for every hour of training, an hour is needed for recovery. Clients almost always think that means stretching for an hour, but recovery comes in many forms. Here are some others.

HYDRATION AND NUTRITION

Water helps carry oxygen to our body cells, which results in a properly functioning system. It also helps remove toxins from the body. When we sweat, we need to compensate for that water loss. The rule of thumb I always give my clients is: if you are thirsty, you are already dehydrated (see p. 27).

When it comes to diet, use our nutrition guide (see pp. 26–27) to find out how best to fuel your workouts, which can make all the difference. Many people don't realize that 80 percent of our immune system lies within our gut, so when it's healthy, we tend to be able to fight off infections faster and more efficiently.

STRETCHING

Stretching helps alleviate tightness and improve flexibility. When performing HIIT exercises, your muscles contract, so it's important to lengthen them by stretching them or they can become unbalanced. Muscle imbalances can lead to stress on the joints and injuries. Also, when muscles are looser, they allow a greater range of motion, which means you'll be carrying out the exercises properly.

The types of stretches include standing quad stretch, lunge with spinal twist, triceps stretch, figure-four stretch, cat stretch, 90/90 stretch, happy baby stretch, lungeing hip flexor stretch, frog stretch, and butterfly stretch.

USING A FOAM ROLLER

Foam rolling is a self-myofascial release (SMR) technique that can help relieve muscle tightness, soreness, and inflammation. In addition, it can also increase your range of motion. So it's well worth including a foam rolling routine in your warm up or cool down.

When foam rolling, start with light pressure. You may find that your muscles are tight and it may be painful to foam roll. Adjust the pressure you apply by reducing the amount of body weight you place on the roller. Roll for about 10 seconds to start with, then work up to 30–60 seconds. If you're new to it, you could ask a professional for help or search online.

FOLLOWING AND CREATING
YOUR OWN ROUTINE

When starting a new training program, you want to create a routine that helps to build a habit. It takes 18 days to build a habit and 66 days for that habit to become automatic, but only two days to break it. Creating a plan of action not only helps develop new healthier habits, but provides a clear line to your overall fitness goals.

LEARNING **THE EXERCISES**

It's important you read things thoroughly before making a start. I always say read things through three times and, if you have any questions, write them down. The chances are that the answers to your questions are in the book. Make sure you understand what is needed from the workout each day and the proper form associated with each exercise. If any of the exercises seem unfamiliar to you, turn to the corresponding page in the step-by-step exercise guide to check how to perform it properly. Because workouts require your focus and concentration, knowing how to perform each move without taking time out to flip through the book during the workout will help keep you in the zone.

WORKOUTS TO BUILD **CARDIOVASCULAR ENDURANCE AND STRENGTH**

Some of the workouts are more cardio-based than others and are designed to get the heart rate up. Increasing cardio-respiratory endurance improves oxygen uptake in the lungs and blood and can help sustain physical activity for longer.

Endurance refers to a person's ability to sustain prolonged exercise. Aerobic endurance is often equated with cardiovascular fitness and requires the circulatory and respiratory systems to supply energy to the muscles to support sustained physical activity. Cardio workouts also help burn calories and fat and boost the metabolism. The main by-product of aerobic metabolism is carbon dioxide gas, which the body disposes of through the blood and out the lungs.

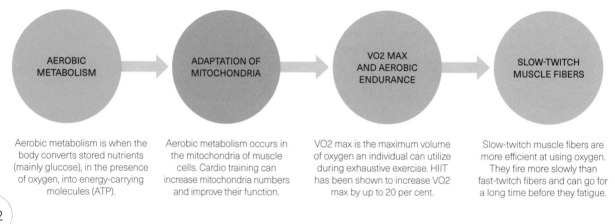

AEROBIC METABOLISM → **ADAPTATION OF MITOCHONDRIA** → **VO2 MAX AND AEROBIC ENDURANCE** → **SLOW-TWITCH MUSCLE FIBERS**

Aerobic metabolism is when the body converts stored nutrients (mainly glucose), in the presence of oxygen, into energy-carrying molecules (ATP).

Aerobic metabolism occurs in the mitochondria of muscle cells. Cardio training can increase mitochondria numbers and improve their function.

VO2 max is the maximum volume of oxygen an individual can utilize during exhaustive exercise. HIIT has been shown to increase VO2 max by up to 20 per cent.

Slow-twitch muscle fibers are more efficient at using oxygen. They fire more slowly than fast-twitch fibers and can go for a long time before they fatigue.

CHOOSING **WEIGHTS**

It's important to start where you are. Make sure you find balance, especially if you're new to HIIT, and choose a weight that will give you a challenge without being too heavy. Understanding how to choose weights (load selection) to fit your structure correctly can impact on the safety and efficacy of your lifting. Start each exercise with a light weight that you know you can lift easily, then progress the weight based on assessment and the desired rep range.

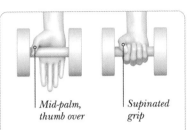

Mid-palm, thumb over | *Supinated grip*

Gripping the weight
You have to grip a weight in a certain way to hold it securely in position and limit soreness in your hands. Common grips are supinated, neutral, and pronated; semi-supinated is halfway between supinated and neutral. Never grip the weight too tightly—this can cause undue stress on your forearms.

Lifting safely
HIIT exercises require focus and constant attention to the body. Focusing on each movement keeps you safe and ensures you can maintain your planned program.

FREE WEIGHTS

Free weights include dumbbells, barbells, and kettlebells. The size of the dumbbells, whether hex, round, or adjustable, is stamped on the weight. Dumbbells are typically found in pairs. Start with a weight that's manageable, one that you can lift with increased reps. As you condition yourself, slowly begin to take the load up to challenge yourself.

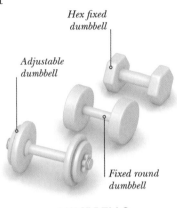

Hex fixed dumbbell

Adjustable dumbbell

Fixed round dumbbell

DUMBBELLS

WORKOUTS TO **TONE AND BUILD STRENGTH**

Exercises that tone and strengthen require resistance from weights, kettlebells, or resistance bands. These types of exercise focus on building muscle mass as well. Anaerobic (meaning without oxygen) exercises include lifting weights or activities that require short bursts of energy. They can be beneficial if you are looking to push through an exercise plateau and meet a new goal, and can also help you maintain muscle mass as you age. Most HIIT training is anaerobic. Anaerobic metabolism produces lactic acid buildup (that's when you feel soreness). Fast-twitch muscles are mostly used. They help you move faster, but for shorter periods. "Twitch" refers to the contraction, or how quickly and often the muscle moves. The body relies more on stored energy to fuel itself for these types of exercise.

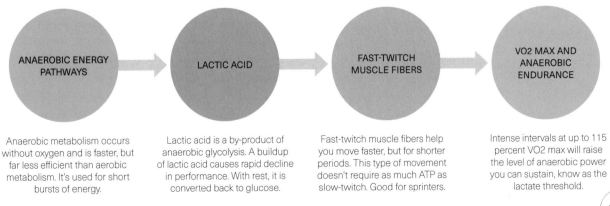

ANAEROBIC ENERGY PATHWAYS	LACTIC ACID	FAST-TWITCH MUSCLE FIBERS	VO2 MAX AND ANAEROBIC ENDURANCE
Anaerobic metabolism occurs without oxygen and is faster, but far less efficient than aerobic metabolism. It's used for short bursts of energy.	Lactic acid is a by-product of anaerobic glycolysis. A buildup of lactic acid causes rapid decline in performance. With rest, it is converted back to glucose.	Fast-twitch muscle fibers help you move faster, but for shorter periods. This type of movement doesn't require as much ATP as slow-twitch. Good for sprinters.	Intense intervals at up to 115 percent VO2 max will raise the level of anaerobic power you can sustain, know as the lactate threshold.

WEEKLY TRAINING
PLANNER

When you make a plan, you're making a plan to succeed. I've put together a six-week progressive plan aimed at helping you jump-start your fitness journey, or improve upon it. Use the routines suggested here or use some of the routines created on pages 199–209.

	MONDAY	TUESDAY	WEDNESDAY	THURSDAY
WEEK 1	**UPPER BODY** Bent-over Wide Row, DB Biceps Curl, Hammer Curl, Alt. Bear Plank Row	**CORE** Bicycle Crunch, Double Crunch Hold with Twist, Scissor Kick, Plank Jack	**LOWER BODY** Sumo Squat, Chair Squat, Crab Walk, Sumo Fly	‖‖‖‖‖‖‖‖‖‖‖‖‖
WEEK 2	Ⓡ	**LOWER BODY** Alt. Snatch, Step Up with DB, Alt. Lateral Lunge, Alt. Toe Tap	**FULL BODY** Bent-over Row + Hammer Curl, Partial Biceps Curl, Tuck Jump, Push Up + Squat	**UPPER BODY** Side-to-side Push Up, DB Bench Press, DB Chest Fly, Triceps Dip
WEEK 3	**FULL BODY** BO Row + Hammer Curl, Sumo Squat + C Curl, B Boy Power Kicks, Bear Plank and Push Up	Ⓡ	**UPPER BODY** DB Bench Press, DB Chest Fly, Triceps Kickback, OH Triceps Extension	**CORE** Swim Plank, Mountain Climber, Rope Pull, Sit Up
WEEK 4	**UPPER BODY** DB Bent-over Row, DB Front Raise, Partial Biceps Curl, Hammer Curl	**CORE** Plank Side-to-side Jump, Bicycle Crunch, Double Crunch, V-Up Around the World	Ⓡ	**LOWER BODY** Single-leg Deadlift, Hamstring Walkout, Calf Raise, Step Up
WEEK 5	**CORE** Scissor Kick, Transverse Abdominal Ball Crunch, Alt. Foot Switch, Double Crunch Hold with Twist	**LOWER BODY** Alt. Back Lunge, Chair Squat, Squat, Skater	**FULL BODY** BW Inverted Shoulder Press, Wide Biceps Curl, Triceps Dip with Toe Touch, Front Curtsy Lunge	Ⓡ
WEEK 6	**FULL BODY** Step Up, Single-leg Deadlift, Crab Walk, Alt. Toe Tap	**UPPER BODY** DB Bent-over Row, DB Biceps Curl, Hammer Curl, Alt. Bear Plank Rows	**CORE** Bicycle Crunch, Double Crunch Hold with Twist, Scissor Kick, Plank Jack	**LOWER BODY** Sumo Squat, Chair Squat, Crab Walk, Sumo Fly

Tracking your progress

There is something very rewarding about crossing out the days you've completed your workouts. It takes about two to three weeks to create a habit, and two days to break one.

The way to succeed is to set a time that you know you can make. Be realistic. If you haven't been working out, then creating a plan where you train every day isn't very realistic. Start with perhaps one day a week, then if you do more, you have exceeded your goal. Decide whether you're more likely to stick with it in the morning, at lunchtime, or in the evening, and be consistent with that time.

To track your progress, you can also take pictures of yourself before you start—front, back, and side. I always say photos don't lie. Take new pictures every two weeks, so you can compare them side-by-side, to see and monitor the changes.

Importance of rest

Rest days are an important part of any exercise routine. Rest days allow the muscles to regenerate tissue and replenish glycogen stores, thereby reducing muscle fatigue and preparing the muscles for the next workout. Rest days also help reduce injuries: overexercising puts repetitive stress and strain on muscles, increasing the risk of injury. Rest days improve performance too. When muscles are allowed to rest, you are better able to perform the next day. Your body needs time to repair and refuel, especially between HIIT workouts.

KEY

- 🔵 Upper body
- 🔵 Core
- 🔵 Lower body
- 🔵 Full body
- ||||| Stretch/Foam roll
- Ⓡ Rest

FRIDAY	SATURDAY	SUNDAY
FULL BODY Football Up and Down, Box Jump, Sumo Squat + Hammer Concentration Curl, Alt. Snatch	**UPPER BODY** Military Press + OH Triceps Extension, DB Lateral Raise, Triceps Kickback, High Plank to Low Plank	**CORE** Dolphin Plank, Low Plank Hold, Leg-reach Alt. Toe Tap, Mountain Climber
||||||||||||||||	**CORE** V Up, Crunch, Sit Up, Bear Plank	**LOWER BODY** Glute Bridge, Butterfly Glute Bridge, Hamstring Walkout, Jump Rope High Knees
LOWER BODY Squat, Single-leg Deadlift, Alt. Curtsy Squat, Squat Jump	||||||||||||||||	**FULL BODY** Transverse Abdominal Ball Crunch, Jack Press, Burpee, High Plank, Ankle Tap + Push Up
FULL BODY Squats + Alt. Kickback, Arnold Press, DB Front Raise, In-and-out Squat Jump	**UPPER BODY** Arnold Press, Banded Upright Row, Hammer Curl, Triceps Push Up	||||||||||||||||
UPPER BODY Neutral-grip DB Shoulder Press, DB Rear Deltoid Fly, Bear Crawl, Push Up	||||||||||||||||	**LOWER BODY** Jump Rope High Knee, Glute Bridge, Frog Jump, Single-leg Deadlift
Ⓡ	**FULL BODY** Football Up and Down, Box Jump, Sumo Squat + Hammer Concentration Curl, Alt. Snatch	||||||||||||||||

WARMING UP AND
COOLING DOWN

To prevent injuries, an adequate warm up and cool down must be a part of your workout. Performing any type of resistance or aerobic training on cold muscles can lead to stress and injuries, so it's vital you get into the habit of warming up the predominant muscles before you start exercising.

> *Performing any type of **resistance on cold** muscles can lead to **stress on the** joints and, potentially, injuries.*

Mobility

According to *Webster's Dictionary*, the definition of "mobility" is "the ability to move your body freely across all planes". Mobility encompasses muscular strength, range of motion, and endurance. Mobility movements are good to add to the beginning of your workout as part of your warm up. They can also serve as a short workout on a rest day. Because mobility increases your range of motion, it incorporates both flexibility and strength, which will help you execute movements more thoroughly and allow you to push harder and jump higher.

WARMING **UP**

A warm up before a HIIT workout activates the muscles, getting them ready to perform and limiting injuries.

The warm up focuses on warming up the cardiovascular system, by raising the body's temperature and increasing blood flow to your muscles. A good solid warm up lasting 5–10 minutes also gets your heart rate up. Depending on the workout ahead, you may sometimes have to lengthen the warm up to get the muscles fully ready to perform. Plyometric movements, for instance, such as box jumps, squat jumps, and burpees, require additional warming up, because this type of movement places a lot of stress on the body. So it's vital that you make thoroughly sure all your muscle groups are ready for that type of strain.

COOLING **DOWN**

Cooling down after a workout allows for recovery, giving you a chance to reduce your heart rate and blood pressure.

The cool down usually takes place after the workout and lasts 5–7 minutes. More often than not, though, the cool down is neglected, rushed through or done haphazardly. I see it all the time, especially in the group-fitness areas of working out. Most people rush out without giving it a second thought. But the truth is, the muscles will remain contracted if you don't lengthen them, keeping them in a continuous state of tension. Think of a rubber band: if you keep pulling and pulling on it without giving it a break, eventually it will snap. It's important to release the tension and return the muscles to a state of relaxation after each workout. Cooling down also helps regulate blood flow.

IMPORTANCE OF STRETCHING

Flexibility is one of the five components of fitness, which is why stretching should be an integral part of every workout program. I always say that anyone over the age of 30 should be operating on a 1:1 ratio; so, for every hour of training, that's one hour of recovery. Stretching is a huge part of that.

Stretching aims to:

• **Decrease muscle stiffness and increase range of motion.** By improving your range of motion, stretching may also ease degeneration in the joints.

• **Reduce risk of injuries.** When muscles are flexible and you make a sudden move, you are less likely to become injured. By increasing the range of motion in a particular joint through stretching, you can decrease the resistance on your body's muscles during various activities.

• **Help relieve post-workout soreness and aches.** When working out, we contract (shorten) the muscles. Stretching after a workout will help lengthen the muscles and alleviate tightness.

• **Improve posture.** Stretching your muscles—especially in the shoulders, lower back, and chest—helps keep the spine in alignment and improves your posture.

• **Help reduce and manage stress.** Well-stretched muscles hold less tension. As a consequence, you feel less stressed.

• **Reduce muscular tension and enhance muscular relaxation.** When muscles spend too much time in a contracted state, they cut off their own circulation, resulting in a limited flow of oxygen and other essential nutrients. Stretching allows muscles to relax by increasing blood flow.

• **Improve overall functional performance and mechanical efficiency.** Flexible joints require less energy to move, so a flexible body improves overall performance by creating energy-efficient movements.

• **Prepare the body for exercise.** When you loosen up the muscles prior to exercising, they are better able to withstand the impact of the movements you do.

• **Promote circulation.** When you relieve tension by stretching the muscles, this increases the blood flow throughout the body, from muscles through to joints. Improved circulation allows more nutrients to be distributed around the body.

• **Decrease the risk of lower-back pain.** If you are having lower-back issues, the chances are the root cause is below it. Tightness in the hamstrings, hip flexors, and several other muscles in the pelvis can pull on the lower back because they are holding tension. Alleviating the tension through stretching can eliminate that pressure.

TRAINING ROUTINE

Your training routine has to include a warm up and a cool down. Your warm up should start with larger muscle groups, then move to more specific body parts. Your warm up should elevate the heart rate, warm the body, and make you sweat.

Warming up
Your options for warming up include:
• A light jog or brisk walk
• High knees, butt kicks, burpees, walkouts
• Push ups
• Swimming
• Jumping jacks

Try a combination of any of the options listed to create a warm up that lasts 5–10 minutes.

Stretching
After your body is warmed up is a good time to add a few stretches, to alleviate tightness and improve flexibility. Doing HIIT exercises, the muscles contract (shorten). It's important to lengthen them by stretching them or they can become unbalanced. Muscle imbalances can lead to stress on the joints and injuries. When the muscles are looser, this allows for greater range of motion, which in turn allows for proper execution of movements.

Types of stretches include:
• Standing quad stretch
• Lunge with spinal twist
• Triceps stretch
• Cat stretch
• Figure four stretch
• 90/90 stretch
• Happy baby stretch
• Frog stretch
• Butterfly stretch
• Lungeing hip flexor stretch
• Lying pectoral stretch

Cooling down
Your cool down could consist of a walk with lower intensity, but it's also the perfect time to add a stretching routine, with some of the exercises above.

WORKOUT **ROUTINES**

The following HIIT routines have been programmed to target specific body parts: lower, upper, core, and full body. The routines have also been devised to be suitable for different levels of experience: beginner, intermediate, and advanced. Choose your experience level, then select a routine and complete a workout to the correct exercise length, rest periods, and number of sets.

HOW **TO USE**

My motto is start where you are, but that's definitely not where you will end. First, complete an assessment of your fitness levels (see pp.188–189) to determine where to begin that journey. Routines are arranged by difficulty, but each can be adjusted to suit your current fitness level by following the guidelines listed in the boxes below. For example, a beginner workout can progress in difficulty by increasing the length of each exercise and/or the number of sets and/or reducing rest between sets or cutting out breaks between exercises. That's what makes this book so priceless, because as you get stronger and faster, you can always find ways to challenge yourself and devise your own routines—variations are almost limitless.

Beginner

Each exercise length is 30 seconds, with a 15-second break between exercises, and 30–60 seconds of recovery between each set.

FOR EXAMPLE

 Exercise for 30 seconds

 Break for 15 seconds

 Rest 30–60 seconds between each set

For a beginner workout, aim to perform 2–3 sets. You can always reduce this if it's too much for you or add a set to make it more challenging.

Intermediate

Each exercise length is 45 seconds, with a 15-second break between exercises, and 30–45 seconds of recovery between each set.

FOR EXAMPLE

Exercise for 45 seconds

Break for 15 seconds

Rest 30–45 seconds between each set

For an intermediate workout, you are looking to increase to 3–4 sets, though reduce this if it's too much or add a set to increase the challenge.

Advanced

Each exercise length is 60 seconds, with no break between exercises, and 30–45 seconds of recovery between each set.

FOR EXAMPLE

 Exercise for 60 seconds

 No break between exercises

 Rest 30–45 seconds between each set

For an advanced workout, aim to complete 4–5 sets, but scale back if you find it too challenging or add a further set to progress.

BEGINNER ROUTINE 1

This routine gets the heart rate up, tones the legs, and strengthens the abdominal muscles. A perfect workout for beginners to improve their cardiovascular endurance and also their muscular strength and endurance.

Beginner

30 secs each exercise, 15-sec break between exercises, for 2–3 sets

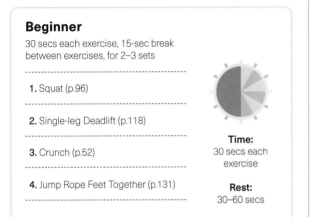

1. Squat (p.96)
2. Single-leg Deadlift (p.118)
3. Crunch (p.52)
4. Jump Rope Feet Together (p.131)

Time:
30 secs each exercise

Rest:
30–60 secs

ROUTINE 2

This is a full body beginner's routine targeting legs and arms, with a cardio component that will push the heart rate up. Perfect for beginners to improve their cardiovascular endurance and also muscular strength and endurance.

Beginner

30 secs each exercise, 15-sec break between exercises, for 2–3 sets

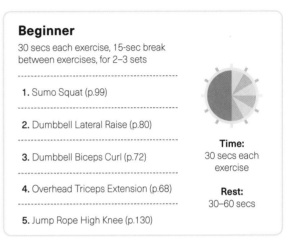

1. Sumo Squat (p.99)
2. Dumbbell Lateral Raise (p.80)
3. Dumbbell Biceps Curl (p.72)
4. Overhead Triceps Extension (p.68)
5. Jump Rope High Knee (p.130)

Time:
30 secs each exercise

Rest:
30–60 secs

ROUTINE 3

This routine targets the legs and the glutes. Here, you will work on toning and strengthening the legs in addition to building those butt muscles. This is perfect for beginners to improve upon muscular strength and endurance.

Beginner

30 secs each exercise, 15-sec break between exercises, for 2–3 sets

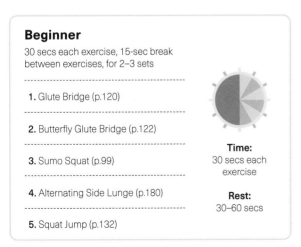

1. Glute Bridge (p.120)
2. Butterfly Glute Bridge (p.122)
3. Sumo Squat (p.99)
4. Alternating Side Lunge (p.180)
5. Squat Jump (p.132)

Time:
30 secs each exercise

Rest:
30–60 secs

ROUTINE 4

This routine focuses on the upper body, specifically toning and strengthening the muscles of the shoulders and triceps. A perfect workout for beginners to improve upon muscular strength and endurance of the upper body.

Beginner

30 secs each exercise, 15-sec break between exercises, for 2–3 sets

1. Military Shoulder Press (p.82)
2. Triceps Kickback (p.70)
3. Triceps Dip (p.71)
4. Inverted Shoulder Press (p.85)
5. Triceps Push-up (p.66)

Time:
30 secs each exercise

Rest:
30–60 secs

ROUTINE **5**

This routine targets the abdominal muscles. Here, you will tone and strengthen the transversus abdominis, rectus abdominis, and external and internal obliques. Perfect for beginners to improve upon your muscular strength and endurance.

Beginner

30 secs each exercise, 15-sec break between exercises, for 2–3 sets

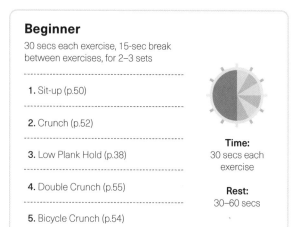

1. Sit-up (p.50)

2. Crunch (p.52)

3. Low Plank Hold (p.38)

4. Double Crunch (p.55)

5. Bicycle Crunch (p.54)

Time:
30 secs each exercise

Rest:
30–60 secs

ROUTINE **6**

This routine targets the workout on the upper body, in particular toning and strengthening muscles of the chest and triceps. A perfect routine for beginners to improve upon their muscular strength and endurance.

Beginner

30 secs each exercise, 15-sec break between exercises, for 2–3 sets

1. Overhead Triceps Extension (p.68)

2. Triceps Kickback (p.70)

3. Dumbbell Bench Press (p.90)

4. Dumbbell Chest Fly (p.92)

5. Side-to-side Push-up (p.67)

Time:
30 secs each exercise

Rest:
30–60 secs

ROUTINE **7**

This routine targets the legs and the deltoids (shoulders). Here, you will work on toning and strengthening the legs in addition to building those shoulders. This is perfect for beginners to improve upon your muscular strength and endurance.

Beginner

30 secs each exercise, 15-sec break between exercises, for 2–3 sets

1. Dumbbell Goblet Squat (p.99)

2. Alternating Back Lunge (p.110)

3. Military Shoulder Press (p.82)

4. Dumbbell Lateral Raise (p.80)

5. Jump Rope (any; pp.130–131)

Time:
30 secs each exercise

Rest:
30–60 secs

ROUTINE **8**

This routine is focused on the upper body, targeting the biceps and abs muscles, with a cardio component to get your heart rate up. Perfect for beginners to improve cardiovascular endurance and also muscular strength and endurance.

Beginner

30 secs each exercise, 15-sec break between exercises, for 2–3 sets

1. Bicycle Crunch (p.54)

2. Hammer Curl (p.75)

3. Wide Biceps Curl (p.74)

4. Partial Biceps Curl (p.75)

5. Alternating Toe Taps (p.116)

Time:
30 secs each exercise

Rest:
30–60 secs

ROUTINE 9

This routine targets the shoulders and abdominal muscles and has a cardio component to it as well in order to get the heart rate up. Perfect for beginners to improve their cardiovascular endurance and also muscular strength and endurance.

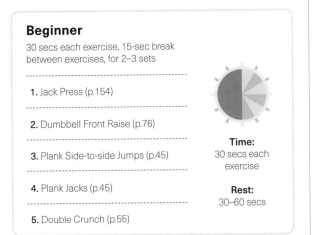

Beginner

30 secs each exercise, 15-sec break between exercises, for 2–3 sets

1. Jack Press (p.154)
2. Dumbbell Front Raise (p.76)
3. Plank Side-to-side Jumps (p.45)
4. Plank Jacks (p.45)
5. Double Crunch (p.55)

Time:
30 secs each exercise

Rest:
30–60 secs

ROUTINE 10

This is an upper body routine targeting the chest and triceps. Here, you will work on toning and strengthening the chest and the triceps. This is perfect for beginners to improve upon your muscular strength and endurance.

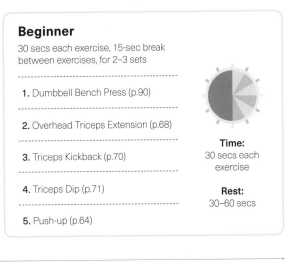

Beginner

30 secs each exercise, 15-sec break between exercises, for 2–3 sets

1. Dumbbell Bench Press (p.90)
2. Overhead Triceps Extension (p.68)
3. Triceps Kickback (p.70)
4. Triceps Dip (p.71)
5. Push-up (p.64)

Time:
30 secs each exercise

Rest:
30–60 secs

ROUTINE 11

This routine works the lower body, targeting the legs and glutes, and has a cardio component to get the heart rate up. Perfect for beginners to improve their cardiovascular endurance, as well as muscular strength and endurance.

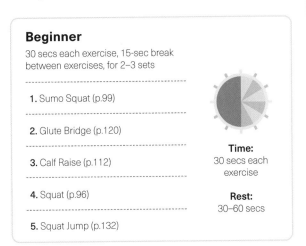

Beginner

30 secs each exercise, 15-sec break between exercises, for 2–3 sets

1. Sumo Squat (p.99)
2. Glute Bridge (p.120)
3. Calf Raise (p.112)
4. Squat (p.96)
5. Squat Jump (p.132)

Time:
30 secs each exercise

Rest:
30–60 secs

ROUTINE 12

This routine targets the legs and has a cardio component to it as well in order to get the heart rate up. A great workout for beginners to improve their cardiovascular endurance, as well as muscular strength and endurance in the legs.

Beginner

30 secs each exercise, 15-sec break between exercises, for 2–3 sets

1. Dumbbell Goblet Squat (p.99)
2. Alternating Curtsy Squat (p.102)
3. Squat (p.96)
4. Single-leg Deadlift (p.118)
5. In-and-out Squat Jump (p.135)

Time:
30 secs each exercise

Rest:
30–60 secs

ROUTINE **13**

This routine is a workout for both the upper and lower body, targeting muscles of the legs and the biceps. This is perfect for beginners to improve their muscular strength and endurance.

Beginner

30 secs each exercise, 15-sec break between exercises, for 2–3 sets

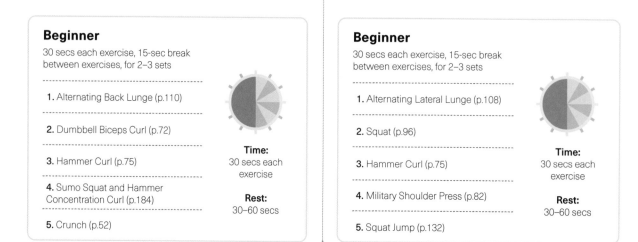

1. Alternating Back Lunge (p.110)

2. Dumbbell Biceps Curl (p.72)

3. Hammer Curl (p.75)

4. Sumo Squat and Hammer Concentration Curl (p.184)

5. Crunch (p.52)

Time:
30 secs each exercise

Rest:
30–60 secs

ROUTINE **14**

This routine targets the legs and biceps, providing a workout for both the upper and lower body. Perfect for beginners to improve their muscular strength and endurance.

Beginner

30 secs each exercise, 15-sec break between exercises, for 2–3 sets

1. Alternating Lateral Lunge (p.108)

2. Squat (p.96)

3. Hammer Curl (p.75)

4. Military Shoulder Press (p.82)

5. Squat Jump (p.132)

Time:
30 secs each exercise

Rest:
30–60 secs

INTERMEDIATE ROUTINE **1**

This routine targets the abs: here, you will tone and strengthen the transversus abdominis, rectus abdominis, and external and internal obliques. This is a perfect intermediate workout to improve upon your muscular strength and endurance.

Intermediate

45 secs each exercise, 15-sec break between exercises, for 3–4 sets

1. Transverse Abdominal Ball Crunch (p.56)

2. V-up Around the World (p.61)

3. Double Crunch Hold with Twist (p.55)

4. Rope Pull (p.54)

5. Scissor Kick (p.60)

Time:
45 secs each exercise

Rest:
30–45 secs

ROUTINE **2**

This routine is an upper body "push-pull" workout that tones and strengthens the muscles of the back and biceps. A perfect intermediate workout to improve your muscular strength and endurance.

Intermediate

45 secs each exercise, 15-sec break between exercises, for 3–4 sets

1. Bent-over Row and Hammer Curl (p.178)

2. Rear Deltoid Fly and Triceps Kickback (p.180)

3. Partial Biceps Curl (p.75)

4. Alternating Row (p.49)

5. Push-up (p.64)

Time:
45 secs each exercise

Rest:
30–45 secs

ROUTINE 3

This routine is another upper body "push-pull" workout, this time toning and strengthening the muscles of the chest and the three muscles that make up the triceps. A perfect intermediate workout to improve upon muscular strength.

Intermediate

45 secs each exercise, 15-sec break between exercises, for 3–4 sets

1. Dumbbell Bench Press (p.90)

2. Dumbbell Chest Fly (p.92)

3. Overhead Triceps Extension (p.68)

4. Triceps Kickback (p.70)

5. Triceps Push-up (p.66)

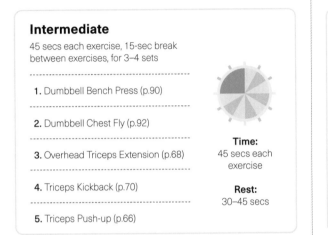

Time:
45 secs each exercise

Rest:
30–45 secs

ROUTINE 4

This routine is an upper body workout targeting the triceps and shoulders. Here, you will work on toning, strengthening, and testing the endurance of your shoulder deltoids and the three muscles that make up the triceps.

Intermediate

45 secs each exercise, 15-sec break between exercises, for 3–4 sets

1. Military Press + Overhead Triceps Extension (p.174)

2. Dumbbell Lateral Raise (p.80)

3. Arnold Press (p.85)

4. Triceps Dip (p.71)

5. Inverted Shoulder Press (p.85)

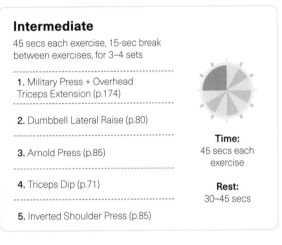

Time:
45 secs each exercise

Rest:
30–45 secs

ROUTINE 5

This is an upper body and core routine that works to tone and strengthen the deltoid muscles and the abdominal muscles. This is a perfect intermediate workout to improve upon your muscular strength and endurance.

Intermediate

45 secs each exercise, 15-sec break between exercises, for 3–4 sets

1. Dumbbell Rear Deltoid Fly (p.86)

2. Banded Upright Row (p.79)

3. Arnold Press (p.85)

4. V-up Around the World (p.61)

5. Bear Crawl (p.150)

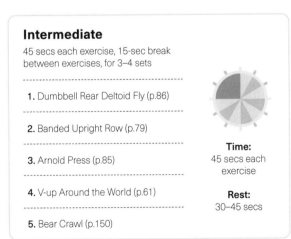

Time:
45 secs each exercise

Rest:
30–45 secs

ROUTINE 6

This routine concentrates on the lower body, working to tone and strengthen all the primary muscles of the leg: quadriceps, hamstrings, and glutes. This routine also has a cardio component to get the heart rate up.

Intermediate

45 secs each exercise, 15-sec break between exercises, for 3–4 sets

1. Squat (p.96)

2. Single-leg Deadlift (p.118)

3. Glute Bridge (p.120)

4. Hamstring Walkout (p.123)

5. Tuck Jump (p.136)

Time:
45 secs each exercise

Rest:
30–45 secs

ROUTINE **7**

This routine targets the lower body, working to tone and strengthen all the primary muscles of the leg: quadriceps, hamstrings, and glutes. There is also a cardio component that will raise your heart rate. Perfect for muscular and cardiovascular endurance.

Intermediate

45 secs each exercise, 15-sec break between exercises, for 3–4 sets

1. Alternating Back Lunge (p.110)

2. Alternating Curtsy Squat (p.102)

3. Walking Lunge with Dumbbells (p.111)

4. Alternating Lateral Lunge (p.108)

5. In-and-out Squat Jump (p.135)

Time:
45 secs each exercise

Rest:
30–45 secs

ROUTINE **8**

This routine is focused on upper body and arms, toning and strengthening muscles of the biceps, shoulders, and triceps. This is a perfect intermediate workout to improve upon your muscular strength and endurance.

Intermediate

45 secs each exercise, 15-sec break between exercises, for 3–4 sets

1. Hammer Curl (p.75)

2. Military Shoulder Press (p.82)

3. Overhead Triceps Extension (p.68)

4. Dumbbell Biceps Curl (p.72)

5. Triceps Dip (p.71)

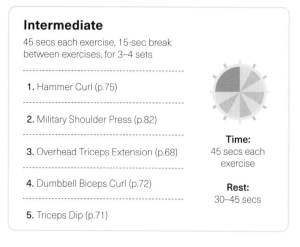

Time:
45 secs each exercise

Rest:
30–45 secs

ROUTINE **9**

This routine works on toning and strengthening the muscles in the leg (quadriceps, hamstrings, and glutes), as well as the shoulders (medial, lateral, rear deltoids). A perfect intermediate workout to improve your muscular strength and endurance.

Intermediate

45 secs each exercise, 15-sec break between exercises, for 3–4 sets

1. Chair Squat (p.98)

2. Squat (p.96)

3. Dumbbell Front Raise (p.76)

4. Dumbbell Lateral Raise (p.80)

5. Inverted Shoulder Press (p.85)

Time:
45 secs each exercise

Rest:
30–45 secs

ROUTINE **10**

This routine targets the abs. Here, you will tone and strengthen the transversus abdominis, rectus abdominis, and external and internal obliques. This is a perfect intermediate workout to improve upon your muscular strength and endurance.

Intermediate

45 secs each exercise, 15-sec break between exercises, for 3–4 sets

1. Double Crunch Hold with Twist (p.55)

2. Bear Plank (p.46)

3. Mountain Climber (p.42)

4. Low Plank Hold (p.38)

5. Plank Jacks (p.45)

Time:
45 secs each exercise

Rest:
30–45 secs

ROUTINE 11

This routine works to tone and strengthen the muscles of the glutes (maximus and minimus) and abdominals. This is a perfect intermediate workout to improve upon muscular strength and endurance.

Intermediate

45 secs each exercise, 15-sec break between exercises, for 3–4 sets

1. Leg Reach Toe Tap (p.48)

2. Hamstring Walkout (p.123)

3. Glute Bridge (p.120)

4. Butterfly Glute Bridge (p.122)

5. Double Crunch (p.55)

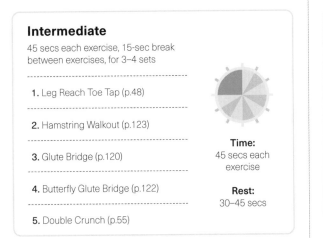

Time:
45 secs each exercise

Rest:
30–45 secs

ROUTINE 12

This lower body routine works to tone and strengthen all the primary muscles of the leg: quadriceps, hamstrings, and glutes. This routine also includes a cardio component designed to raise your heart rate.

Intermediate

45 secs each exercise, 15-sec break between exercises, for 3–4 sets

1. Squat (p.96)

2. Single-leg Deadlift (p.118)

3. Sumo Squat (p.99)

4. Jump Rope Butt Kick (p.131)

5. In-and-out Squat Jump (p.135)

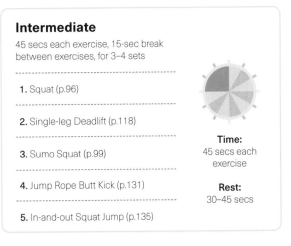

Time:
45 secs each exercise

Rest:
30–45 secs

ROUTINE 13

This routine is a focused on the lower body. Here, you will work on toning and strengthening all the primary muscles of the leg: quadriceps, hamstrings, and glutes. The routine also has a cardio component.

Intermediate

45 secs each exercise, 15-sec break between exercises, for 3–4 sets

1. Sumo Squat (p.99)

2. Squat (p.96)

3. Single-leg Deadlift (p.118)

4. Crab Walk (p.104)

5. Frog Jump (p.134)

Time:
45 secs each exercise

Rest:
30–45 secs

ROUTINE 14

This routine targets the lower body, especially the main leg muscles—quadriceps, hamstrings, and glutes—which it aims to tone and strengthen. It also has a cardio component and is a perfect for improving muscular and cardiovascular endurance.

Intermediate

45 secs each exercise, 15-sec break between exercises, for 3–4 sets

1. Chair Squat (p.98)

2. Squat (p.96)

3. Crab Walk (p.104)

4. Alternating Curtsy Squat (p.102)

5. Squat Jump (p.132)

Time:
45 secs each exercise

Rest:
30–45 secs

ADVANCED ROUTINE 1

This routine is an upper body "push-pull" workout to tone and strengthen the back and biceps, with an added power component to get the heart rate up. This is a challenging advanced workout to improve upon your strength, endurance, and agility.

Advanced

60 secs each exercise, NO break between exercises, for 4–5 sets

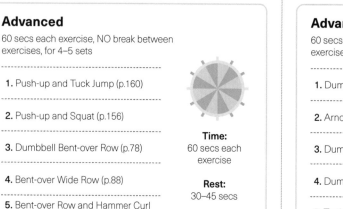

1. Push-up and Tuck Jump (p.160)

2. Push-up and Squat (p.156)

3. Dumbbell Bent-over Row (p.78)

4. Bent-over Wide Row (p.88)

5. Bent-over Row and Hammer Curl (p.178)

Time:
60 secs each exercise

Rest:
30–45 secs

ROUTINE 2

This full body plyometric routine tones and strengthens the muscles of the back (latissimus dorsi), deltoids (medial, lateral, rear), and legs (quadriceps). A challenging advanced workout to improve strength, endurance, and agility.

Advanced

60 secs each exercise, NO break between exercises, for 4–5 sets

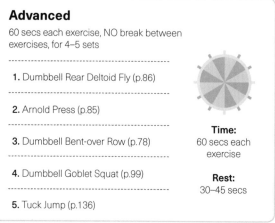

1. Dumbbell Rear Deltoid Fly (p.86)

2. Arnold Press (p.85)

3. Dumbbell Bent-over Row (p.78)

4. Dumbbell Goblet Squat (p.99)

5. Tuck Jump (p.136)

Time:
60 secs each exercise

Rest:
30–45 secs

ROUTINE 3

This is a leg- and back-focused routine that will tone and strengthen the quadriceps, hamstrings, glutes, and back. A challenging workout to improve muscular strength and endurance, which also includes a cardio component.

Advanced

60 secs each exercise, NO break between exercises, for 4–5 sets

1. Squat (p.96)

2. Single-leg Deadlift (p.118)

3. Dumbbell Pullover (p.89)

4. Banded Upright Row (p.79)

5. In-and-out Squat Jump (p.135)

Time:
60 secs each exercise

Rest:
30–45 secs

ROUTINE 4

This routine is an upper body workout targeting mainly the chest. Here, you will work on toning and strengthening the pectoralis major and minor. This is a challenging advanced workout to improve upon your muscular strength and endurance.

Advanced

60 secs each exercise, NO break between exercises, for 4–5 sets

1. Dumbbell Bench Press (p.90)

2. Dumbbell Chest Fly (p.92)

3. Diamond Push-up (p.67)

4. Dumbbell Bench Press (p.90)

5. Bear Plank and Push-up (p.164)

Time:
60 secs each exercise

Rest:
30–45 secs

ROUTINE 5

This routine is an upper body workout focusing primarily on the arms. Here, you will work on toning and strengthening the muscles of the biceps and triceps. This is a challenging workout to improve upon your muscular strength and endurance.

Advanced

60 secs each exercise, NO break between exercises, for 4–5 sets

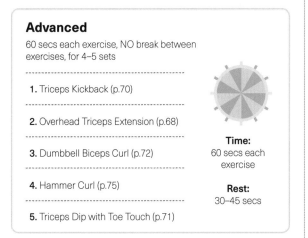

1. Triceps Kickback (p.70)

2. Overhead Triceps Extension (p.68)

3. Dumbbell Biceps Curl (p.72)

4. Hammer Curl (p.75)

5. Triceps Dip with Toe Touch (p.71)

Time:
60 secs each exercise

Rest:
30–45 secs

ROUTINE 6

This lower body routine targets all the primary muscles of the leg: quadriceps, hamstrings, and glutes. This routine has a plyometric (power) component to it as well in order to get the heart rate up and improve agility.

Advanced

60 secs each exercise, NO break between exercises, for 4–5 sets

1. Alternating Toe Tap (p.116)

2. Alternating Curtsy Squat (p.102)

3. Alternating Lateral Lunge (p.108)

4. Squat (p.96)

5. Box Jump (p.138)

Time:
60 secs each exercise

Rest:
30–45 secs

ROUTINE 7

This routine is focused on the lower body, specifically targeting the glutes and hamstrings to tone and strengthen these muscles. This routine also has a plyometric (power) component to get the heart rate up and improve agility.

Advanced

60 secs each exercise, NO break between exercises, for 4–5 sets

1. Glute Bridge (p.120)

2. Hamstring Walkout (p.123)

3. Single-leg Deadlift (p.118)

4. Step-up with Dumbbells (p.114)

5. Tuck Jump (p.136)

Time:
60 secs each exercise

Rest:
30–45 secs

ROUTINE 8

This upper body routine works on toning and strengthening primarily the back muscles: rhomboid minor, rhomboid major, trapezius, and latissimus dorsi. It also has a plyometric component for cardiovascular endurance and improved agility.

Advanced

60 secs each exercise, NO break between exercises, for 4–5 sets

1. Push-up and Tuck Jump (p.160)

2. Dumbbell Bent-over Row (p.78)

3. Dumbbell Rear Deltoid Fly (p.86)

4. Bent-over Wide Row (p.88)

5. Push-up (p.64)

Time:
60 secs each exercise

Rest:
30–45 secs

ROUTINE **9**

This routine strengthens muscles of the lower body and shoulders: quadriceps, hamstrings, and deltoids. There is also a plyometric component to get the heart rate up and improve agility. A workout to improve strength and athletic performance.

Advanced
60 secs each exercise, NO break between exercises, for 4–5 sets

1. Alternating Snatch (p.106)

2. Military Shoulder Press (p.82)

3. Arnold Press (p.85)

4. Crab Walk (p.104)

5. Box Jump (p.138)

Time:
60 secs each exercise

Rest:
30–45 secs

ROUTINE **10**

This routine targets the core and secondarily the shoulders. Here, you will work on toning and strengthening the muscles in the deltoids and the abdominal muscles. A challenging workout to improve strength, endurance, and mobility.

Advanced
60 secs each exercise, NO break between exercises, for 4–5 sets

1. B Boy Power Kicks (p.172)

2. High Plank, Ankle Tap, and Push-up (p.168)

3. Plank Side-to-side Jumps (p.45)

4. Swim Plank (p.40)

5. Bear Plank (p.46)

Time:
60 secs each exercise

Rest:
30–45 secs

ROUTINE **11**

Here, you will work on toning and strengthening the muscles in the chest—pectoralis major and minor—and also the triceps. This is a challenging workout to improve muscular strength and endurance, cardiovascular endurance and agility.

Advanced
60 secs each exercise, NO break between exercises, for 4–5 sets

1. Dumbbell Bench Press (p.90)

2. Dumbbell Chest Fly (p.92)

3. Overhead Triceps Extension (p.68)

4. Triceps Kickback (p.70)

5. Triceps Push-up and Tuck Jump (p.160 with variation p. 66)

Time:
60 secs each exercise

Rest:
30–45 secs

ROUTINE **12**

This routine works on toning and strengthening the primary muscles of the glutes, and also has a plyometric component to get the heart rate up and improve agility. A challenging workout to improve strength, endurance, and athletic performance.

Advanced
60 secs each exercise, NO break between exercises, for 4–5 sets

1. Glute Bridge (p.120)

2. Butterfly Glute Bridge (p.122)

3. Jump Rope Butt Kick (p.131)

4. Donkey Kick (p.48)

5. Frog Jump (p.134)

Time:
60 secs each exercise

Rest:
30–45 secs

ROUTINE 13

A core routine to tone and strengthen the muscles of the abdominals. This is a challenging advanced workout to improve upon your muscular strength and endurance, cardiovascular endurance, mobility, and athletic performance.

Advanced

60 secs each exercise, NO break between exercises, for 4–5 sets

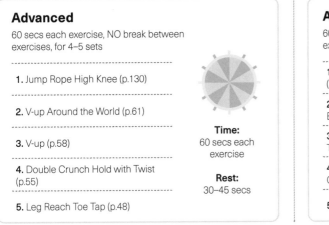

1. Jump Rope High Knee (p.130)

2. V-up Around the World (p.61)

3. V-up (p.58)

4. Double Crunch Hold with Twist (p.55)

5. Leg Reach Toe Tap (p.48)

Time:
60 secs each exercise

Rest:
30–45 secs

ROUTINE 14

A full body workout targeting the back, biceps, shoulders, triceps, legs, and chest, with a plyometric component to get the heart rate up and improve agility. A challenging workout to improve strength, endurance, and athletic performance.

Advanced

60 secs each exercise, NO break between exercises, for 4–5 sets

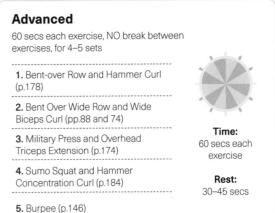

1. Bent-over Row and Hammer Curl (p.178)

2. Bent Over Wide Row and Wide Biceps Curl (pp.88 and 74)

3. Military Press and Overhead Triceps Extension (p.174)

4. Sumo Squat and Hammer Concentration Curl (p.184)

5. Burpee (p.146)

Time:
60 secs each exercise

Rest:
30–45 secs

> *Everyone's body responds differently to training. Focus within on your own effort and intensity and **follow each** **routine** closely. The more you put into your workouts, the faster you will see results and a **change** **in your** body.*

GLOSSARY

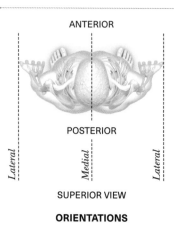

ANTERIOR

POSTERIOR

Lateral *Medial* *Lateral*

SUPERIOR VIEW

ORIENTATIONS

Aerobic Relating to, or being, activity that increases the body's demand for oxygen, resulting in a temporary increase in breathing and heart rate.

Aerobic respiration The way your body creates energy through conversion in the presence of oxygen of glucose into adenosine triphosphate (ATP).

Abdominals A group of muscles in the torso consisting of the rectus abdominis, the external abdominal obliques, the internal abdominal obliques, and the transverse abdominis; commonly know as "the abs."

Abduction The action of moving a limb away from the midline of the body.

Actin A protein that interacts with myosin to cause muscle contraction.

Adduction The action of moving a limb towards the midline of the body.

Adductors A group of muscles that are used to draw the thighs toward the midline, consisting of the adductor longus, the adductor brevis, the adductor magnus, the pectineus, and the gracilis.

Agonist The muscle in an antagonistic muscle pair that contracts as the other one relaxes or lengthens.

Anaerobic Relating to, or being, activity that breaks down glucose in the body without using oxygen; anaerobic means "without oxygen." Anaerobic exercise is more intense but shorter in duration than aerobic exercise.

Anaerobic respiration The creation of energy through the conversion of glucose in the absence of oxygen.

Antagonist The muscle in an antagonistic muscle pair that relaxes or lengthens as the other muscle contracts.

Anterior Situated at the front.

ATP (adenosine triphosphate) An organic compound and hydrotrope that provides energy to drive many processes in living cells, such as muscle contraction, nerve impulse propagation, condensate dissolution, and chemical synthesis.

Bilateral On both sides of the body simultaneously.

BMI (body mass index) A value derived by dividing a person's body mass by the square of their body height, expressed in units of kg/m^2.

Body fat composition A way of describing the percentages of fat, bone, water, and muscle in human bodies. Because muscular tissue takes up less space in the body than fat tissue, body composition, as well as weight, determines leanness.

Carbohydrates Naturally occurring chemical substances that contain carbon, hydrogen, and oxygen; they are the primary source of energy when stored in the body.

Cardiovascular system The biological system that permits blood to circulate and transport nutrients (such as amino acids and electrolytes), oxygen, carbon dioxide, hormones, and blood cells to and from the cells in the body to provide nourishment, fight diseases, stabilize temperature and pH, and maintain homeostasis. Also called the vascular system.

Coactivation When multiple muscles activate simultaneously.

Compound movements Any movement in which you're using more than one muscle group at a time.

Concentric contractions A type of muscle activation that causes tension on your muscle as it shortens.

Deadlift An exercise that involves extending at the knees and/or hips to lift a weight from the ground.

Deltoid A muscle of the shoulder.

Dumbbell A type of exercise equipment consisting of a short bar with weights at either end; usually used in a pair.

Eccentric contractions A type of muscle activation in which a force applied to the muscle exceeds the momentary force produced by the muscle itself, causing the muscle-tendon system to lenghen while contracting.

Endurance The ability of an organism to exert itself and remain active for a long period of time, as well as its ability to resist, withstand, recover from, and have immunity to trauma, wounds, or fatigue. It is usually used in aerobic or anaerobic exercise.

EPOC (excess post-exercise oxygen consumption) A measurable increased rate of oxygen intake following strenuous activity.

Extension A movement that increases the angle of a joint.

Fast-twitch muscle fibers Muscle fibers that support short, quick bursts of energy, such as sprinting or weightlifting. These muscles operate anaerobically, so have few blood vessels and mitochondria (unlike slow-twitch muscles).

Fat A nutrient with several necessary functions in the body, including protecting internal organs and nerves and assisting the absorption of vitamins.

Fitness assessment A series of tests that measure and monitor your physical fitness level by assessing the five components of physical fitness: cardiovascular endurance, muscular strength, muscular endurance, flexibility, and body composition.

Flexion A movement that decreases the angle of a joint.

Frontal plane Any vertical plane (or imaginary line) that divides the body into ventral (front) and dorsal (back) sections. It is one of the three main planes of the body used to describe the location of body parts in relation to each other axis. Also called the coronal plane.

Glucose A simple sugar that is the body's preferred energy source.

Glutes A group of muscles in the buttocks consisting of the gluteus maximus, the gluteus medius, and the gluteus minimus.

Glycogen A substance deposited in bodily tissues as a store of carbohydrates. It is a polysaccharide which forms glucose on hydrolysis.

Glycolysis A process in which glucose (sugar) is partially broken down by cells in enzyme reactions that do not need oxygen. Glycolysis is one method that cells use to produce energy.

HIIT (high-intensity interval training): A cardiovascular exercise strategy alternating short periods of intense anaerobic strength exercise and aerobic cardio with recovery periods, until you are too exhausted to continue.

Hypertrophy The enlargement of an organ or tissue from the increase in size of its cells; used especially in relation to muscle growth.

Isometric contractions A type of muscle activation that involves the static contraction of a muscle without any visible movement in the angle of the joint.

Isotonic muscle contractions A type of muscle activation where tension remains the same while the muscle's length changes. Isotonic contractions differ from isokinetic contractions in that in isokinetic contractions the muscle speed remains constant.

Lactic acid A colorless, syrupy organic acid formed in sour milk and also produced in the muscle tissues during strenuous anaerobic exercise.

Lateral Situated at the side(s).

Mechanical stress The mechanism of converting mechanical energy into chemical signals; also known as mechanotransduction.

Mechanical tension A type of force that tries to stretch a material. During strength training, muscles experience stretching forces when they try to shorten, but are resisted when they do so.

Medial Situated in the middle.

Median plane A vertical plane (or imaginary line) that bisects the body vertically through the midline marked by the navel, dividing the body exactly into the left and right sides. Also called the mid-sagittal plane.

Metabolic rate Metabolism per unit of time, especially as estimated by food consumption, energy released as heat, or oxygen used in metabolic processes.

Metabolic stress A physiological process that occurs during exercise in response to low energy; it leads to metabolite accumulation (lactate, phosphate inorganic [Pi] and ions of hydrogen [H+]) in muscle cells.

Metabolism The chemical reactions in the body's cells that change food into energy. Our bodies need this energy to do everything from moving to thinking to growing. Specific proteins in the body control the chemical reactions of metabolism.

Mitochondria An organelle found in large numbers in most cells, in which the biochemical processes of respiration and energy production occur. It has a double membrane, the inner layer being folded inward to form layers (cristae).

Mobility The ability of a joint to move through a range of motion.

Muscular system The biological system responsible for movement of the body, posture, and blood circulation. The muscular systems in vertebrates are controlled through the nervous system, although some muscles can be completely autonomous.

Myosin-A A protein that interacts with actin to cause muscle contraction.

Nervous system The biological system responsible for sensation and movement. It is made up of the central nervous system and the peripheral nervous system: The brain and the spinal cord are the central nervous system. The nerves that go through the whole body make up the peripheral nervous system.

Neutral grip A way of holding a weight, cable, etc., where the wrists are not rotated, so the palms face each other.

Neutral spine The position of optimal load distribution for the spine; maintains the natural curves of the spine.

Neurochemistry The study of chemicals that control and influence the physiology of the nervous system, including neurotransmitters and other molecules such as psychopharmaceuticals and neuropeptides.

Neurogenesis The process in which neurons are generated from neural stem cells in adults. This process differs from prenatal neurogenesis.

Neuroplasticity The ability of neural networks in the brain to change through growth and reorganization. These changes range from individual neuron pathways making new connections, to systematic adjustments like cortical remapping. Also known as neural plasticity or brain plasticity.

Parasagittal plane Any plane (or imaginary line) parallel to the sagittal and median planes.

Pectorals (pecs) A group of muscles in the chest consisting of the pectoralis major and the pectoralis minor.

Physiology The branch of biology that deals with the normal functions of living organisms and their parts.

Plyometric Exercises in which muscles exert maximum force in short intervals of time, with the goal of increasing power. Also known as jump training or plyos.

Posterior Situated at the back.

Pronated Of a hand, foot, or limb: turned or held so that the palm or sole is facing downwards or inwards.

Prone Lying on your front.

Protein A molecule made up of amino acids; dietary protein is necessary for life and bodily maintenance.

Quadriceps (quads) A group of muscles in the thigh consisting of the rectus femoris, the vastus medialis, the vastus lateralis, and the vastus intermedius.

Range of motion The extent of possible movement of a joint.

Repetitions (reps) The number of times you complete a single exercise before taking a rest or a break.

Resistance An external force against which muscles contract, such as weight.

Respiratory system The biological system consisting of specific organs and structures used for gas exchange in animals and plants. The anatomy and physiology that make this happen vary depending on the size of the organism, the environment in which it lives, and its evolutionary history.

Rhomboids A group of muscles in the upper back consisting of the rhomboid minor and the rhomboid major.

RIR (reps in reserve) A measure of the difficulty of a set that refers to the number of further reps that the person could have performed before giving way to fatigue.

Sagittal plane Any plane (or imaginary line) that divides the body into right and left parts. The plane may be in the center of the body and split it into two halves, or away from the midline and split it into unequal parts. Also known as the longitudinal plane.

Satellite cells A type of body cell within muscle fibers (between the sarcolemma and basal lamina) that is typically in a quiescent state. Following exercise, satellite cells become activated, and start to proliferate.

Sets Several reps of different exercises completed in a row. It is a common workout strategy to do a planned number of sets of each exercise, with time built in for a short rest between these sets.

Skeletal muscles Muscles which are connected to the skeleton to form part of the mechanical system that moves the limbs and other parts of the body.

Skeletal muscle fibers Cylindrical muscle cells. An individual skeletal muscle may be made up of hundreds, or even thousands, of muscle fibers bundled together and wrapped in a connective tissue covering.

Skeletal system The body's central framework. It consists of bones and connective tissue, including cartilage, tendons, and ligaments.

Slow-twitch muscle fibers Muscle fibers that contract slowly, but keep going for a long time. These muscles are good for endurance activities like long-distance running or cycling, as they can work for a long time without getting tired.

Strength The amount of force a muscle or muscle group can produce.

Strength endurance The ability of muscles to support load continuously over time. Also known as muscular endurance.

Stress Mechanical, metabolic, or psychological demands placed on the body.

Striated muscle Muscle tissue in which the contractile fibrils in the cells are aligned in parallel bundles, so that their different regions form stripes visible in a microscope. Muscles of this type are attached to the skeleton by tendons and are under voluntary control.

Superficial (of muscles) Closer to the skin.

Supersets Two exercises performed back-to-back, optionally followed by a short rest. This effectively doubles the amount of work you are doing, whilst keeping the recovery periods the same as they are when you complete individual exercises.

Supinated Of a hand, foot, or limb: turned or held so that the palm or sole is facing upwards or outwards.

Supine Lying on your back.

Tempo The rhythm in which exercises are performed during sets.

Tendon A flexible but inelastic cord of strong fibrous collagen tissue attaching a muscle to a bone.

Transverse plane A plane (or imaginary line) that divides the body into superior (upper) and inferior (lower) parts. It is perpendicular to the coronal plane and sagittal plane. It is one of the planes of the body used to describe the location of body parts in relation to each other.

Unilateral One side of the body.

VO2 max The maximum rate of oxygen consumption measured during incremental exercise (that is, exercise of increasing intensity).

INDEX

A

abdominals (abs) 22, 36, 38, 39, 40, 46, 50, 58, 64, 88, 90, 102, 132
  external abdominal obliques 22, 39, 50, 51, 54, 60
  internal abdominal obliques 22, 50, 51, 54, 60
  rectus abdominis 22, 38, 39, 40, 50, 52, 54, 56, 117, 120
  transversus abdominis 22, 38, 40, 50, 54, 56, 122
  see also core exercises
abduction 32, 33
adduction 32, 33
adductors 22
  adductor brevis 22
  adductor longus 22
  adductor magnus 22, 23
  gracilis 22
  pectineus 22
adenosine triphosphate see ATP
aerobic endurance 192
aerobic respiration 12, 13, 14, 27
"afterburn effect" see EPOC (Excess Post-exercise Oxygen Consumption)
agonist muscles 18, 19
alternating back lunge 110

alternating curtsy squat 102
alternating foot switch 44
alternating lateral lunge 108–111
alternating row 49
alternating single-leg glute bridge 122
alternating snatch 106–107
alternating toe tap 116–117
Alzheimer's 15
amino acids 26
anaerobic efficiency 6
anaerobic glycolysis 12, 13, 193
anaerobic phosphagen 12, 13
anaerobic respiration 12, 13, 17
anconeus 66
angiogenesis 25
ankle
  dorsiflexion 33
  plantarflexion 33
ankle dorsiflexors 22
  extensor digitorum longus 22
  extensor hallucis longus 22
  tibialis anterior 22
ankle plantor flexors 23
  calf muscles 23
  tibialis posterior 23
antagonist muscles 18, 19

anxiety 6
Arnold press 84, 85
ATP (adenosine triphosphate) 12, 13, 15, 16, 20, 192
  accessing 12, 13
  aerobic synthesis of 15

B

B boy power kicks 172–173
banded upright row 79
basal metabolic rate (BMR) 12, 16
bear crawl 150–151
bear plank 46–49
  alternating row 49
  bear plank and push-up 164–167
  donkey kick 48
  leg-reach toe tap 48
bench press: dumbbell bench press 90–91
bent-over row and hammer curl 178–179
bent-over wide row 88
biceps brachii 18, 19, 22, 72, 78, 92
  see also upper body exercises
biceps curl
  dumbbell biceps curl 18–19, 72–75
  hammer curl 75
  partial biceps curl 75

wide biceps curl 74
biceps femoris 23
bicycle crunch 54
blood
  blood flow 10, 24, 27
  circulation 11, 14, 30, 197
  deoxygenated blood 14
  hemoglobin 14, 17
blood sugar levels 16
blood vessel growth 25
BMI see body mass index
BMR see basal metabolic rate
body fat composition: calculation 189
body mass index (BMI) 189
body temperature
  core 16
  regulation 27
bodyweight inverted shoulder press 85
bodyweight resistance 10
box jump 138–139
brachialis 18, 22, 72, 74
brachioradialis 18, 22, 72, 74
brain cells 24, 25
brain chemistry 24–25
  brain–body connection 24, 25, 30, 31
  neurochemistry 25
  neurogenesis 25
  neuroplasticity 25
breathing 30, 31
bridge

alternating single-leg glute bridge 122
butterfly glute bridge 122
glute bridge 120–123
hamstring walkout 123
burpee 146–149
football up and down 142–145
butterfly glute bridge 122

C

calf muscles 23
gastrocnemius 23
soleus 23
calf raise 112–113
calorie burning 6, 10, 11, 16, 149
capillaries 14
carbohydrates 12, 26, 27
carbon dioxide 13, 14, 31, 192
cardiac muscle 22
cardiovascular disease 15
cardiovascular fitness 6, 11, 14–15
blood circulation 11, 14, 30, 197
mitochondrial function 15
scoring 15
workouts for 192
see also plyometric exercises; total body exercises

cardiovascular system 14, 30, 196
cells
brain cells 24, 25
cytoplasm 13
mitochondria 10, 13, 15, 21, 192
myocytes 21
red blood cells 17
satellite cells 21
cervical extensors 23
splenius capitis 23
splenius cervicis 23
chair squat 98
chemistry
cognitive function 24
see also brain
chemistry
collagen 23
concentric muscle contractions 18, 19, 39
connective tissue 18, 20, 22, 23
hypertrophy 21
cooling down 196, 197
Cooper, Dr Ken 15
Cooper test 15
coordination 11
core exercises 34–61
alternating foot switch 44
alternating row 49
B boy power kicks 172–173
bear crawl 150–151
bear plank 46–49
bicycle crunch 54
crunch 52–55

dolphin plank 39
donkey kick 48
double crunch 55
double crunch hold with twist 55
high plank to low plank 36–39
leg-reach toe tap 48
low plank hold 38
low-impact plank 38
mountain climber 42–45
plank jacks 45
plank side-to-side jumps 45
rope pull 54
scissor kick 60
side V-up 60
sit-up 50–51
swim plank 40–41
transverse abdominal ball crunch 56–57
V-up 58–61
V-up around the world 61
see also total body exercises
cortisol 25
crab touch see triceps dip with toe touch
crab walk 104–105
crunch 52–55
bicycle crunch 54
double crunch 55
double crunch hold with twist 55
rope pull 54
transverse abdominal

ball crunch 56–57
cytoplasm 13

D

deoxygenated blood 14
deadlift, single-leg 118–119
deltoids 23, 64, 76, 78, 79, 80, 82, 84, 85, 86, 88, 90, 92
see also upper body exercises
dementia 15, 24
depression 6, 24
diabetes 15
diamond push-up 67
dolphin plank 39
donkey kick 48
dopamine 25
dorsiflexion 33
double crunch 55
double crunch hold with twist 55
dumbbells 31, 193
alternating snatch 106–107
Arnold press 84, 85
banded upright row 79
bent-over row and hammer curl 178–179
bent-over wide row 88
dumbbell bench press 90–91
dumbbell bent-over row 78

dumbbell biceps curl 18–19, 72–75
dumbbell chest fly 92–93
dumbbell front raise 76–79
dumbbell goblet squat 99
dumbbell lateral raise 80–81
dumbbell pullovers 89
dumbbell rear deltoid fly 86–89
hammer curl 75
jack press 154–155
military press and overhead triceps extension 174–177
military shoulder press 82–85
neutral-grip dumbbell shoulder press 84
overhead triceps extension 68–71
partial biceps curl 75
single-leg deadlift 118–119
step-up with dumbbells 114–115
sumo squat and hammer concentration curl 184–185
triceps kickback 70
walking lunge with dumbbells 111
wide biceps curl 74

E

eccentric muscle contractions 18
elbow
  extension 32
  flexion 32
elbow flexors 22
  biceps brachii *see* biceps brachii
  brachialis 18, 22, 72, 74
  brachioradialis 18, 22, 72, 74
endorphins 24
endurance 192
energy conversion 12–17
energy release 15
EPOC (Excess Post-exercise Oxygen Consumption) 16–17, 27, 190
  EPOC recovery time 17
equipment 31
erector spinae 23, 40, 78, 120
Excess Post-exercise Oxygen Consumption *see* EPOC
exercise balls 31
  transverse abdominal ball crunch 56–57
exercise terminology 31, 32–33
  see also Glossary pp. 210–213
extension 32, 33
extensor digitorum longus 22
extensor hallucis longus 22
external abdominal obliques 22, 39, 50, 51, 54, 60

F

fast-twitch muscle fibers 20, 193
fats 12, 26
femur 98
fight-or-flight scenario 12
fitness assessment 188
flexibility 197
  see also stretching
flexion 32, 33
foam-rolling 191
football up and down 142–145
frog jump 134
front curtsy lunge 110
front plank with rotation *see* mountain climber
frontal plane 32

G

gastrocnemius 23, 112
glucose 12, 13, 15, 16, 26, 192, 193
gluteals (glutes) 11, 23, 36, 50, 96, 102, 104, 106, 108, 118, 120, 132
  gluteus maximus 11, 23, 46, 122, 123
  gluteus medius 23, 46, 123
  gluteus minimus 23, 123
  see also lower body exercises
glycogen 16, 17, 21, 26, 27
glycolysis 12, 13, 193
gracilis 22

H

hemoglobin 14, 17
hammer curl 75
  bent-over row and hammer curl 178–179
  sumo squat and hammer concentration curl 184–185
hamstring walkout 123
hamstrings 23, 58, 96, 100, 106, 122, 132
  biceps femoris 23
  semimembranosus 23
  semitendinosus 23
  see also lower body exercises
heart rate 11, 12, 125
high knee 128–131
  jump rope butt kick 131
  jump rope feet together 131
  jump rope high knee 130
high plank to low plank 36–39
HIIT (high-intensity interval training)
  benefits of 6, 10–11
  brain gains 24–25
  cardiovascular fitness, improving 14–15
  energy systems 12–13
  Excess Post-exercise Oxygen Consumption (EPOC) 16–17, 27, 190
  and muscle growth 20–21
  muscular anatomy 22–23
  nutrition 26–27

physiology of 8–27
hip
  abduction 32
  adduction 32
  extension 32
  external rotation 32
  flexion 32
  internal rotation 32
hip extensors 23
  adductor magnus 23
  gluteals see gluteals
  hamstrings see
    hamstrings
hip flexors 22, 50
  adductors 22
  iliacus 22
  iliopsoas 22
  psoas major 22, 46
  rectus femoris 22, 98
  sartorius 22
hippocampus 24, 25
homeostatic balance 16
hormones 24–25, 26
humerus 66
hydration 27, 191
hypertrophy see muscle
  growth

I

iliacus 22
iliocostalis 78
iliopsoas 22
immune support 10
in-and-out squat jump
  135
inflammation 10
injuries 196, 197

intercostal muscles 22
intermyofibrillar
  mitochondria 15
internal abdominal
  obliques 22, 50, 51, 54,
  60
isometric contractions
  18, 19
isometric exercises 46,
  47, 164
isotonic contractions 18
  concentric contractions
    18, 19, 39
  eccentric contractions
    18

J

jack press 154–155
jumps
  box jump 138–139
  frog jump 134
  in-and-out squat jump
    135
  jump rope butt kick 131
  jump rope feet together
    131
  jump rope high knee
    130
  plank jacks 45
  plank side-to-side
    jumps 45
  push-up and tuck jump
    160–163
  single-leg forward jump
    140–141
  squat jump 132–135
  sumo squat jump 135
  tuck jump 136–137

K

kettlebells 31
  overhead triceps
    extension 68–71
  see also dumbbells
kicks
  B boy power kicks
    172–173
  donkey kick 48
  scissor kick 60
  squat and alternating
    kickback 103
knee
  alignment 111
  extension 33
  flexion 33

L

lactate 13, 17
  lactate threshold 193
lactic acid 193
lactic acidosis 13, 193
latissimus dorsi 23, 78, 88,
  89
leg-reach toe tap 48
levator scapulae 23
longissimus thoracis 78
low plank hold 38
low-impact plank 38
lower body exercises
  94–123
  alternating back lunge
    110
  alternating curtsy squat
    102
  alternating lateral lunge
    108–111
  alternating single-leg

glute bridge 122
  alternating snatch
    106–107
  alternating toe tap
    116–117
  butterfly glute bridge
    122
  calf raise 112–113
  chair squat 98
  crab walk 104–105
  dumbbell goblet squat
    99
  front curtsy lunge 110
  glute bridge 120–123
  hamstring walkout 123
  right and left split squat
    100–103
  single-leg deadlift
    118–119
  squat 96–99
  squat and alternating
    kickback 103
  step-up with dumbbells
    114–115
  sumo squat and sumo
    fly 99
  walking lunge with
    dumbbells 111
  see also total body
    exercises
lower-back pain 120, 197
lunge
  alternating back lunge
    110
  alternating lateral lunge
    108–111
  front curtsy lunge 110
  walking lunge with
    dumbbells 111

M

macronutrients 16, 26
meals, skipping 16
mechanical fatigue 20
mechanical tension 20
metabolic rate 6, 12, 16
  basal metabolic rate (BMR) 12, 16
metabolism 10
  aerobic metabolism 12, 13, 192
  anaerobic metabolism 12, 13, 193
  metabolic efficiency 16
  metabolic stress 20
micronutrients 26
military press and overhead triceps extension 174–177
military shoulder press 82–85
mind–body connection 24, 25, 30, 31
mitochondria 10, 13, 15, 21, 192
  intermyofibrillar mitochondria 15
  mitochondrial function 15
  subsarcolemmal mitochondria 15
mobility 196
mood, boosting 24, 25
mountain climber 42–45
  alternating foot switch 44
  plank jacks 45
  plank side-to-side jumps 45
multifidus 78

muscle ache 13
muscle coactivation 19
muscle contractions 14, 16, 18–19, 22
  concentric contractions 18, 19, 39
  eccentric contractions 18
  isometric contraction 18, 19
  isotonic contraction 18
  voluntary/involuntary 18
muscle damage 20, 21
muscle fibers 10, 15, 21, 22
  fast-twitch muscle fibers 20, 193
  slow-twitch muscle fibers 20, 192
muscle growth 11, 20–21, 193
  growth stimuli 20
muscle hypertrophy see muscle growth
muscle mass decline 21
muscle protein 20, 21, 22
  muscle protein synthesis 20, 27
muscular anatomy 22–23
myocytes 21
myofibrillar hypertrophy 21
myofibrils 15, 20, 22
myoglobin 17
myotendinous junction 23
myotubes 21

N

nausea 13

nervous system 30
neurochemistry 25
neurogenesis 25
neurons 25
neuroplasticity 25
neurotransmitters 25
neurotrophins 24, 25
neutral spine 32
neutral-grip dumbbell shoulder press 84
norepinephrine 25
nutrition 26–27, 191
  balanced diet 26
  hydration 27
  macronutrients 16, 26
  micronutrients 26
  pre- and post-workout nutrition 27
  skipping meals 16

O

obliques see external abdominal obliques; internal abdominal obliques
overhead triceps extension 68–71
  triceps kickback 70
overhydration 27
overtraining 190
oxygen 12, 13, 14, 31, 192
  brain function and 24, 25
  Excess Post-exercise Oxygen Consumption (EPOC) 16–17, 27, 190
  oxygen deficit 17
  VO2 max 6, 15, 17, 192

P

partial biceps curl 75
patella 98
PCr see phosphocreatine
pectineus 22
pectoral muscles (pecs) 64, 76, 82, 90
  pectoralis major 22, 88, 89
  pectoralis minor 22
phosphagen 12, 13
phosphocreatine (PCr) 13
piriformis 23
planes of motion 32
plank
  alternating foot switch 44
  bear plank 46–49
  bear plank and push-up 164–167
  dolphin plank 39
  front plant with rotation see mountain climber
  high plank, ankle tap, and push-up 168–171
  high plank to low plank 36–39
  low plank hold 38
  low-impact plank 38
  plank jacks 45
  plank side-to-side jumps 45
  swim plank 40–41
plantarflexion 33
plyometric exercises 124–151, 196
  bear crawl 150–151
  box jump 138–139
  burpee 146–149
  football up and down

142–145
frog jump 134
high knee 128–131
in-and-out squat jump
135
jump rope butt kick 131
jump rope feet together
131
jump rope high knee
130
push-up and tuck jump
160–163
single-leg forward jump
140–141
skater 126–127
squat jump 132–135
sumo squat jump 135
tuck jump 136–137
posture 11, 197
pronation 32
protein (muscle) 20, 21, 22
protein (nutrient) 12, 26,
27
psoas major 22, 46
push-up 64–67
bear plank and push-up
164–167
diamond push-up 67
high plank, ankle tap,
and push-up 168–171
push-up and squat
156–159
push-up and tuck jump
160–163
side-to-side push-up 67
triceps push-up 66
pyruvate 13
pyruvic acid 15

Q

quadratus lumborum 78
quadriceps (quads) 22,
36, 42, 46, 58, 96, 98,
106, 108, 114
rectus femoris 22
vastus intermedius 22
vastus lateralis 22
vastus medialis 22
see also lower body
exercises

R

range of motion 32–33
ankle 33
elbow 32
hip 32
knee 33
shoulder 33
spine 32
wrist 32
rear deltoid fly and
triceps kickback
180–183
rectus abdominis 22, 38,
39, 40, 50, 52, 54, 56,
117, 120
rectus femoris 22, 98
red blood cells 17
repetition (reps) 31
reps in reserve (RIR)
191
resistance bands 31
banded upright row 79
overhead triceps
extension 68–71
resistance training
benefits 11

respiration 12, 30, 31
aerobic respiration 12,
13, 14, 27
anaerobic respiration 12,
13, 17
rest days 195
rhomboids 23, 78, 88
rhomboid major 23
rhomboid minor 23
right and left split squat
100–103
roll mats 31
rope pull 54
rotation 32, 33
rotatores 78
routines 31, 198–209
see also training
programs
rowing
alternating row 49
banded upright row 79
bent-over row and
hammer curl 178–179
bent-over wide row 88
dumbbell bent-over row
78

S

saccharides 26
sagittal plane 32
sarcolemma 15
sarcomeres 21
sarcoplasm 15
sarcoplasmic fluid 20
sarcoplasmic hypertrophy
21
sarcoplasmic reticulum
21
sartorius 22

satellite cells 21
scapular retraction 79, 86,
180
scissor kick 60
sedentary lifestyle 21
self-myofascial release
(SMR) 191
semimembranosus 23
semispinalis capitis 78
semispinalis thoracis 78
semitendinosus 23
serotonin 25
serratus anterior 64, 80
serratus posterior 23
sets 31
adding 191
shoulder
abduction 33
adduction 33
extension 33
external rotation 33
flexion 33
internal rotation 33
shoulder press
Arnold press 84, 85
bodyweight inverted
shoulder press 85
military shoulder press
82–85
neutral-grip dumbbell
shoulder press 84
side flexion 32
side V-up 60
side-to-side push-up 67
single-arm dumbbell
snatch see alternating
snatch
single-leg deadlift 118–
119
single-leg forward jump
140–141

sit-up 50–51
"six pack" 38, 52, 56
skater 126–127
skeletal muscle 20, 21, 22–23
sleep quality 24
slow-twitch muscle fibers 20, 192
smooth muscle 22
sodium depletion 27
soleus 23
spinal extensors 23, 78
  cervical extensors 23
  erector spinae 23, 119
  transversospinales 23
spinalis thoracis 78
spine
  extension 32
  flexion 32
  rotation 32
  side flexion 32
splenius capitis 23
splenius cervicis 23
squats 96–99
  alternating curtsy squat 102
  alternating snatch 106–107
  burpee 146–149
  chair squat 98
  crab walk 104–105
  dumbbell goblet squat 99
  frog jump 134
  in-and-out squat jump 135
  push-up and squat 156–159
  right and left split squat 100–103
  squat and alternating

kickback 103
  squat jump 132–135
  sumo squat and hammer concentration curl 184–185
  sumo squat jump 135
  sumo squat and sumo fly 99
  step-up with dumbbells 114–115
stomach pain 13
stress, mental 10
  stress reduction 24
stretching 191, 197
  aims of 197
  types of stretches 197
striated muscle 22
sumo squat and hammer concentration curl 184–185
sumo squat jump 135
sumo squat and sumo fly 99
supination 32
supraspinatus 80
sweating 27, 191
swim plank 40–41
synapses 25
synergist muscles 18, 19

T

t-tubules 21
tendons 18, 22, 23
tensor fasciae latae 11
tibialis anterior 22
tibialis posterior 23
toe taps
  alternating toe tap 116–117

leg-reach toe tap 48
total body exercises 152–185
  alternating snatch 106–107
  B boy power kicks 172–173
  bear plank and push-up 164–167
  bent-over row and hammer curl 178–179
  box jump 138–139
  burpee 146–149
  dolphin plank 39
  dumbbell goblet squat 99
  high plank, ankle tap, and push-up 168–171
  jack press 154–155
  military press and overhead triceps extension 174–177
  push-up and squat 156–159
  push-up and tuck jump 160–163
  rear deltoid fly and triceps kickback 180–183
  sumo squat and hammer concentration curl 184–185
  swim plank 40–41
  triceps dip with toe touch 71
training programs 186–209
  advanced programs 189, 190, 198, 206–209
  beginner programs 189, 190, 198, 199–202

for cardiovascular endurance and strength 192
  common mistakes 30
  cooling down 196, 197
  correct exercise execution 30
  creating a routine 192–193
  equipment 31
  exercise terminology 31, 32–33
  fitness assessment 188
  intermediate programs 189, 190, 198, 202–205
  learning the exercises 192
  "main" exercises and "variations" 30
  overtraining 190
  progress, tracking 15, 195
  progressing training 191
  recovery 191
  rest days 195
  routines 198–209
  for tone and strength building 193
  training frequency 190
  warming up 196, 197
  weekly training planner 194–195
  weights, choosing 193
transverse abdominal ball crunch 56–57
transverse plane 32
transversospinales 23
transversus abdominis 22, 38, 40, 50, 54, 56, 122
trapezius 23, 78, 82, 86, 88

upper trapezius 79
triceps (triceps brachii) 18, 19, 23, 36, 64, 66, 82, 90, 92
    see also upper body exercises
triceps dip 71
triceps dip with toe touch 71
triceps kickback 70
triceps push-up 66
tuck jump 136–137

U

ulna 66
upper body exercises 62–93
    Arnold press 84, 85
    banded upright row 79
    bent-over wide row 88
    bodyweight inverted shoulder press 85
    diamond push-up 67
    dumbbell bench press 90–91
    dumbbell bent-over row 78
    dumbbell biceps curl 72–75
    dumbbell chest fly 92–93
    dumbbell front raise 76–79
    dumbbell lateral raise 80–81
    dumbbell pullovers 89
    dumbbell rear deltoid fly 86–89
    hammer curl 75

military shoulder press 82–85
neutral-grip dumbbell shoulder press 84
overhead triceps extension 68–71
partial biceps curl 75
push-up 64–67
side-to-side push-up 67
triceps dip 71
triceps dip with toe touch 71
triceps kickback 70
triceps push-up 66
wide biceps curl 74
    see also total body exercises

V

V-up 58–61
    scissor kick 60
    side V-up 60
    V-up around the world 61
vastus intermedius 22, 98
vastus lateralis 11, 22, 98
vastus medialis 22, 98
vitamins and minerals 26
VO2 12, 17, 193
    Cooper test 15
    VO2 max 6, 15, 17, 192

W

walking lunge with dumbbells 111
warming up 196, 197

waste products 14, 27
    see also carbon dioxide
weekly training planner 194–195
weight loss 11
weights
    choosing 193
    free weights 193
    gripping 193
    lifting safely 193
    see also dumbbells
wide biceps curl 74
workouts see training programs
wrist
    pronation 32
    supination 32

BIBLIOGRAPHY

http://www.unm.edu/~lkravitz/Article%20folder/Metabolism.pdf

http://www.biobreeders.com/images/Nutrition_and_Metabolism.pdf

Kirkendall DT, Garrett WE. Function and biomechanics of tendons. Scand J Med Sci Sports. 1997 Apr;7(2):62-6.

https://www.registerednursing.org/teas/musculoskeletal-muscular-system/

El-Sayes J, Harasym D, Turco CV, Locke MB, Nelson AJ. Exercise-Induced Neuroplasticity: A Mechanistic Model and Prospects for Promoting Plasticity. Neuroscientist. 2019 Feb

Schoenfeld, Brad J The Mechanisms of Muscle Hypertrophy and Their Application to Resistance Training, Journal of Strength and Conditioning Research: October 2010 - Volume 24 - Issue 10 - p 2857-2872

https://health.gov/dietaryguidelines/2015/guidelines/appendix-7/

National Science Teaching Association: "How Does the Human Body Turn Food Into Useful Energy?"

American Council on Exercise: "Muscle Fiber Types: Fast-Twitch vs. Slow-Twitch"

International Sports Sciences Association: "Aerobic vs. Anaerobic: How Do Workouts Change the Body?"

World Journal of Cardiology: "Aerobic vs Anaerobic Exercise Training Effects on the Cardiovascular System"

Health.gov: "Dietary Guidelines for Americans, 2015-2020: Appendix 7. Nutritional Goals for Age-Sex Groups Based on Dietary Reference Intakes and Dietary Guidelines Recommendations"

https://www.sciencedirect.com/topics/biochemistry-genetics-and-molecular-biology/phosphagen

https://www.sciencedirect.com/topics/medicine-and-dentistry/creatine-phosphate

PubMed: Effects of Plyometric Training on Muscle-Activation Strategies and Performance in Female Athletes

PubMed: The Efficacy and Safety of Lower-Limb Plyometric Training in Older Adults: A Systematic Review

Boyle, M. New Functional Training for Sports, 2nd ed. Champaign, IL. Human Kinetics; 2016.

Clark, MA, et al. NASM Essentials of Personal Fitness Training 6th ed. Burlington, MA. Jones & Bartlett Learning; 2018.

McGill, EA, Montel, I. NASM Essentials of Sports Performance Training, 2nd Edition. Burlington, MA. Jones & Bartlett Learning; 2019.

Chu, DA. Jumping Into Plyometrics 2nd ed. Champaign, IL: Human Kinetics; 1998.

Chu, D and Myers, GD. Plyometrics: Dynamic Strength and Explosive Power. Champaign, IL. Human Kinetics (2013).

EXOS Phase 3 Performance Mentorship manual. San Diego. July 27-30, 2015

Fleck, SJ, Kraemer, WJ. Designing Resistance Training Programs 2nd ed. Champaign, IL: Human Kinetics; 1997.

Rose, DJ. Fall Proof! A Comprehensive Balance and Mobility Training Program. Champaign, IL: Human Kinetics; 2003.

Yessis, M. Explosive Running: Using the Science of Kinesiology to Improve Your Performance (1st Edition). Columbus, OH. McGraw-Hill Companies. (2000).

American College of Sports Medicine. ACSM's Guidelines for Exercise Testing and Prescription. 9th ed. Philadelphia (PA): Lippincott Williams and Wilkins; 2013. pp. 19–38.

https://journals.lww.com/acsm-healthfitness/fulltext/2014/09000/high_intensity_interval_training__a_review_of.5.aspx#O20-5-2

Gibala MJ, Little JP, Macdonald MJ, Hawley JA. Physiological adaptations to low-volume, high-intensity interval training in health and disease. J Physiol. 2012; 590: 1077–84.

Gibala MJ, McGee SL. Metabolic adaptations to short-term high-intensity interval training: a little pain for a lot of gain? Exerc Sport Sci Rev. 2008; 36: 58–63.

Guiraud T, Nigam A, Gremeaux V, Meyer P, Juneau M, Bosquet L. High-intensity interval training in cardiac rehabilitation. Sports Med. 2012; 42: 587–605.

Hegerud J, Hoydal K, Wang E, et al Aerobic high-intensity intervals improve VO2 max more than moderate training. Med Sci Sports Exerc. 2007; 39: 665–71

Jung M, Little J. Taking a HIIT for physical activity: is interval training viable for improving health. In: Paper presented at the American College of Sports Medicine Annual Meeting: Indianapolis (IN). American College of Sports Medicine; 2013.

Wewege M, van den Berg R, Ward RE, Keech A. The effects of high-intensity interval training vs. moderate-intensity continuous training on body composition in overweight and obese adults: a systematic review and meta-analysis. Obes Rev. 2017 Jun;18(6):635

Nicolò A, Girardi M. The physiology of interval training: a new target to HIIT. J Physiol. 2016;594(24):7169-7170

Milioni F, Zagatto A, Barbieri R, et al. Energy Systems Contribution in the Running-based Anaerobic Sprint Test. International Journal of Sports Medicine. 2017;38(03):226-232

Abe T, Loenneke JP, Fahs CA, Rossow LM, Thiebaud RS, Bemben MG. Exercise intensity and muscle hypertrophy in blood flow-restricted limbs and non-restricted muscles: a brief review. Clin Physiol Funct Imaging 32: 247–252, 2012.

Aebersold R, Mann M. Mass-spectrometric exploration of proteome structure and function. Nature 537: 347–355, 2016.

Agergaard J, Bülow J, Jensen JK, Reitelseder S, Drummond MJ, Schjerling P, Scheike T, Serena A, Holm L. Light-load resistance exercise increases muscle protein synthesis and hypertrophy signaling in elderly men. Am J Physiol Endocrinol Metab 312

Allen DG, Lamb GD, Westerblad H. Skeletal muscle fatigue: cellular mechanisms. Physiol Rev88: 287–332, 2008

American College of Sports Medicine. American College of Sports Medicine position stand. Progression models in resistance training for healthy adults. Med Sci Sports Exerc 41: 687–708, 2009

Callahan MJ, Parr EB, Hawley JA, Camera DM. Can High-Intensity Interval Training Promote Skeletal Muscle Anabolism? Sports Med. 2021 Mar;51(3):405-421

Børsheim E, Bahr R. Effect of exercise intensity, duration and mode on post-exercise oxygen consumption. Sports Med. 2003; 33(14): 1037-60

LaForgia J, Withers RT, Gore CJ. Effects of exercise intensity and duration on the excess post-exercise oxygen consumption. J Sports Sci. 2006 Dec;24(12):1247-64

Baker, J. S., McCormick, M. C., & Robergs, R. A. (2010). Interaction among Skeletal Muscle Metabolic Energy Systems during Intense Exercise. Journal of nutrition and metabolism, 2010

Mukund K, Subramaniam S. Skeletal muscle: A review of molecular structure and function, in health and disease. Wiley Interdiscip Rev Syst Biol Med. 2020;12(1)

Hasan, Tabinda. (2019). Science of Muscle Growth: Making muscle.

McNeill Alexander R. Energetics and optimization of human walking and running: the 2000 Raymond Pearl memorial lecture. Am J Hum Biol. 2002 Sep-Oct;14(5):641-8.

Arias P, Espinosa N, Robles-García V, Cao R, Cudeiro J. Antagonist muscle co-activation during straight walking and its relation to kinematics: insight from young, elderly and Parkinson's disease. Brain Res. 2012 May 21;1455:124-31

Scott, Christopher. "Misconceptions about Aerobic and Anaerobic Energy Expenditure." Journal of the International Society of Sports Nutrition vol. 2,2 32-7. 9 Dec. 2005

Alberts B, Johnson A, Lewis J, et al. Molecular Biology of the Cell. 4th edition. New York: Garland Science; 2002. How Cells Obtain Energy from Food.

de Freitas MC, Gerosa-Neto J, Zanchi NE, Lira FS, Rossi FE. Role of metabolic stress for enhancing muscle adaptations: Practical applications. World J Methodol. 2017 Jun 26;7(2):46-54.

Van Horren B, et al. Do we need a cool-down after exercise? A narrative review of the psychophysiological effects and the effects on performance, injuries and the long-term adaptive response. Sports Medicine. 2018;48:1575.

https://www.ncbi.nlm.nih.gov/pmc/articles/PMC6548056/
https://www.ncbi.nlm.nih.gov/pmc/articles/PMC4180747/
https://www.ncbi.nlm.nih.gov/pmc/articles/PMC5554572/
https://www.ncbi.nlm.nih.gov/pubmed/2150579
https://pubmed.ncbi.nlm.nih.gov/21997449/
https://pubmed.ncbi.nlm.nih.gov/28394829/
https://pubmed.ncbi.nlm.nih.gov/29781941/
https://pubmed.ncbi.nlm.nih.gov/26102260/
https://pubmed.ncbi.nlm.nih.gov/18438258/
https://pubmed.ncbi.nlm.nih.gov/14599232/
https://pubmed.ncbi.nlm.nih.gov/17101527/

ABOUT THE AUTHOR

Ingrid S Clay, is a Celebrity Personal Trainer, Master HIIT Group Fitness Instructor, Competitive Bodybuilder, and a Plant-Based Chef with over a decade of professional experience in fitness and wellness. She deeply understands the direct impact fitness has on individual success and happiness.

From Lafayette Louisiana, Ingrid graduated with a degree in Physics from Xavier University of LA; a degree in Electrical Engineering from North Carolina A&T University; and received her MBA in International Marketing from Simmons School of Management. Her science background influences the way she views fitness and wellness as a whole.

After putting on weight while working full-time and attending night school, Ingrid went back to the food and fitness basics and what she knew best: science. She created her own diet and exercise regime, incorporating mostly HIIT-based weighted workouts. She enrolled in Simmons School of Management's Entrepreneurial Program, left her corporate job, and began building her own wellness company full-time. She started working in fitness, working on accreditations, and studying from trainers who had been in the industry for years. And ISC Wellness was born.

Ingrid has traveled the world training, coaching, and cooking. She has been featured in Well + Good, Essence, Livestrong, Fabletics, and PopSugar Fit. Ingrid owns and operates ISC Wellness and has an App that features both live and pre-recorded workouts. Ingrid is a Lululemon Brand Ambassador and currently Director of Fitness at CAMP in Los Angeles, California. In addition, Ingrid volunteers weekly with the Watts Empowerment Center, providing health and wellness to kids in the underserved Watts Community.

"Writing this book has been a dream come true. I truly hope you enjoy it and reach your goals! Best of luck! My passion for inspiring others and helping others be their best selves is my purpose; there is no ending to what we are becoming."

For more information on Ingrid:
www.ingridsclay.com

To donate to the Watts Empowerment Center:
www.youthmentor.org/givehope

ACKNOWLEDGMENTS

Author's acknowledgments
I'd like to first thank my family for always supporting me. Especially my mom. You have always been my number one cheerleader and fan. It's because of your support I am able to fly so high!

I'd also like to thank my clients, from personal training to everyone I've met in studios where I've coached. I've learned so much from you and I am constantly inspired by you. Thank you for letting me be a part of your journeys.

Thanks to Chuck Norman, for teaching me the ropes and for helping me always find the fun in Fitness.

Lastly I'd like to also thank Fitness. It was a form of meditation for me when I needed it the most. It brought me back in so many ways, and it made me stronger from the inside out.

Publisher's acknowledgments
Dorling Kindersley would like to thank Myriam Megharbi for picture research, Marie Lorimer for indexing, Guy Leopold for proofreading, and Holly Kyte for editorial assistance.

Picture credits
The publisher would like to thank the following for their kind permission to reproduce their photographs:
(Key: a-above; b-below/bottom; c-center; f-far; l-left; r-right; t-top)

10 Science Photo Library: Professors P.M. Motta, P.M. Andrews, K.R. Porter & J. Vial (br). **14 Science Photo Library**: Ikelos GmbH / Dr. Christopher B. Jackson (cra). **15 Science Photo Library**: CNRI (br). **22 Science Photo Library:** Professors P.M. Motta, P.M. Andrews, K.R. Porter & J. Vial (bl). **25 Science Photo Library**: Thomas Deerinck, NCMIR (ca).

All other images © **Dorling Kindersley**
For further information see: **www.dkimages.com**